Poisonous Snakes
of the World

Poisonous Snakes
of the World

BY

U.S. Department of the Navy
(Bureau of Medicine and Surgery)

DOVER PUBLICATIONS, INC.
NEW YORK

Published in Canada by General Publishing Company, Ltd., 30 Lesmill Road, Don Mills, Toronto, Ontario.

Published in the United Kingdom by Constable and Company, Ltd., 3 The Lanchesters, 162-164 Fulham Palace Road, London W6 9ER.

This Dover edition, first published in 1991, is an unabridged republication of the U.S. Department of the Navy, Bureau of Medicine and Surgery, manual NAVMED P-5099, *Poisonous Snakes of the World: A Manual for Use by U.S. Amphibious Forces,* originally published by the U.S. Government Printing Office, Washington, D.C., [1966]. The eight color plates, originally occupying pages 205–212, have here been transferred to an insert following page 12.

Manufactured in the United States of America
Dover Publications, Inc., 31 East 2nd Street, Mineola, N.Y. 11501

Library of Congress Cataloging-in-Publication Data

Poisonous snakes of the world / by U.S. Department of the Navy, Bureau of Medicine and Surgery.
     p.   cm.
  Reprint. Originally published: Washington, D.C. : U.S. Govt. Print. Off., 1966.
Originally published in series: NAVMED P.
  Includes bibliographical references and index.
  ISBN 0-486-26629-X (pbk.)
  1. Poisonous snakes. 2. Poisonous snakes—Venom—Toxicology. I. United States. Navy Dept. Bureau of Medicine and Surgery.
QL666.O6P56 1991
597.96—dc20
                                                90-20200
                                                  CIP

FOREWORD

The first edition of the ONI 3-62, *Poisonous Snakes of the World*, was published on 30 June 1962 under the auspices of the Office of Naval Intelligence of the Office of the Chief of Naval Operations. The widespread interest generated by this publication and the increasing commitments of Navy and Marine forces throughout the world, particularly in Southeast Asia, have served to emphasize the need for a more authoritative manual that can be used for training and in support of military operations. The Bureau of Medicine and Surgery, having recognized this necessity, assumed the responsibility for a more definitive and comprehensive up-to-date presentation of the problems relating to venomous snakes.

Commander Granville M. Moore, MSC, USN, was appointed as coordinator and principal editor to work with a committee of eminent herpetologists, selected by the American Society of Ichthyologists and Herpetologists, to revise the manual. The Bureau of Medicine and Surgery gratefully acknowledges the important contributions made by this committee and the services provided by the New York Zoological Society.

First aid procedures in cases of snakebite as described herein are approved by the Bureau of Medicine and Surgery and the specific treatment set forth represents the official policy of this Bureau at the time of publication.

This manual is recommended for use by all ships, stations, and commands in need of authoritative information about snakes and snakebites.

R. B. Brown
Vice Admiral, MC
United States Navy
Surgeon General and Chief,
Bureau of Medicine and Surgery

PREFACE

This revision has been made with the assistance of a committee appointed by the American Society of Ichthyologists and Herpetologists. The committee consisted of Dr. Herndon G. Dowling, Dr. Sherman A. Minton, Jr. (chairman), and Dr. Findlay E. Russell. The text has been largely rewritten, however, many of the original illustrations have been retained.

This manual is intended to serve as a training aid and as an identification guide to the most widely distributed species of dangerously venomous snakes. Geographic distribution of all currently recognized species of venomous snakes is presented in tabular form. Information on habitat and biology of important snake species has been provided.

First aid procedures in case of snakebite and suggestions for the definitive medical management of the snakebite victim are presented. There is a table of world sources of antivenins.

The manuscript for the text of this manual was submitted for publication on 1 November 1965. A few additions have been made during the editing and proofreading of the text but most of the included information is as of the date of submission.

ACKNOWLEDGMENTS

In revising this manual, the members of the Committee gratefully acknowledge the advice and assistance of the following, each of whom is a recognized authority in some aspect of biology or medicine: C. A. Ahuga, Steven C. Anderson, H. D. Baernstein, Charles M. Bogert, F. W. Buess, W. Leslie Burger, Roger Conant, Carl Gans, Joseph F. Gennaro, Jr., Itzchak Gilboa, Alice C. Grandison, Laurence M. Klauber, Robert E. Kuntz, Alan E. Leviton, Hymen Marx, Samuel B. McDowell, K. A. C. Powell, George B. Rabb, H. Alistair Reid, Janis A. Roze, the late F. A. Shannon, Harold Voris, John E. Werler, and Eric Worrell.

Personnel of the Medical Photography Division, Naval Medical School, National Naval Medical Center, prepared most of the line illustrations and furnished some of the photographs. Dr. T. E. Reed, Director of the National Zoological Park, kindly made available certain specimens for photography. The index was prepared by Itzchak Gilboa.

S. A. M., Jr.
H. G. D.
F. E. R.

TABLE OF CONTENTS

CHAPTER VII

CHAPTER VIII

CHAPTER IX

---

    * To facilitate use of this chapter as a reference work, a separate table of contents has been provided.
    ** To facilitate use of this chapter as a reference work, separate tables of contents have been provided and placed at the beginning of each geographic section.

LIST OF TABLES

LIST OF COLOR PLATES
(following page 12)

Poisonous Snakes
of the World

Chapter I

GENERAL INFORMATION

Personnel of the U.S. Navy and Marine Corps may find themselves stationed or visiting in many parts of the world, particularly the countries bordering the oceans. In some of these countries, snakebite is a significant public health hazard. The risk of being bitten increases during amphibious operations, especially in tropical and subtropical regions. During such operations the natural habitat of venomous snakes may be disturbed so that exposure to them is markedly increased.

American military forces have never experienced casualty rates from snake venom poisoning sufficiently high to jeopardize the outcome of an operation. However, the threat of snakebite may create a morale problem sufficient to delay an operation or cause unnecessary fear during its execution. While snakebite has been rare and fatalities therefrom have been even more uncommon in the military forces, it does constitute a medical emergency requiring immediate attention and considerable judgment in management.

This manual is designed to facilitate identification of the major groups (genera) of poisonous snakes and to identify the most dangerous species. It is not practical to by-pass the specialized terminology of herpetology completely, but herpetological terms are avoided whenever possible.

Those that are used are defined in the glossary or are evident from examination of the figures.

Geographic definitions of regions discussed are provided because of differences in the use of such words as *Middle East, Southeast Asia, Near East, et cetera*. Snakes found in more than one region are listed in each.

A second aim of the manual is to give suggestions for preventing snakebite, and a third aim is to indicate practical first aid measures should snakebite occur. Principles and procedures for medical management of snake venom poisoning are discussed, but it is not a purpose of this manual to evaluate all of the varied and sometimes conflicting therapeutic regimens that have appeared in the medical literature.

A list of general references is included at the end of the manual, and most chapters and sections are followed by a list of specific references. A space for notes will be found at the end of most chapters and sections. This may be used for additional references and information gained under local conditions.

The index has been prepared as a major source of information. Many local or vernacular names are found *only* in the index, where they are referred to the scientific name of the species.

Chapter II

PRECAUTIONS TO AVOID SNAKEBITE

The best way to keep from being bitten by snakes is to avoid them. However, since there is little choice in a duty assignment, there are certain precautions to be taken in "snake country." In such areas it is advisable to carry a snakebite first aid kit. Snakebite Kit, Suction (FSN 6545-952-5325), may be ordered through the Armed Forces Supply Agency. When such kits are not available, the following items can be substituted: an antiseptic, a razor or sharp knife, a piece of rubber tubing or similar item to be used as a tourniquet, and any device capable of providing suction. A 10 ml. syringe with needle, a vial of physiologic saline and two vials of adrenalin should also be carried for use in administering horse serum sensitivity tests (see p. 16).

Reminders

When in snake infested country it is important to:

1. *Remember that snakes are probably more afraid of humans than humans are of snakes.* Given the chance snakes will usually retreat to avoid an encounter.

2. *Learn to recognize the poisonous snakes in the area of operation.* Avoid killing harmless snakes.

3. *Avoid walking around after dark.* Many venomous snakes are nocturnal and will travel at night far beyond the distances they may venture during the day. If you must walk at night be sure to wear boots.

4. *Remember that snakes in general avoid direct sunlight*, and that they are most active at moderate temperatures.

5. *Avoid caves, open tombs, and known snake den areas.* Snakes live in areas which afford protection and which may be frequented by other small animals. They may be found in considerable numbers in caves and open tombs during the hibernation period which in most snakes extends from fall until early spring. They may also seek out these same areas during the summer months.

6. *Remember that poisonous snakes may be found at high altitudes*, and that they can climb trees and fences.

7. *Walk on clear paths as much as possible.* Avoid tall grass and areas of heavy underbrush or ground covering. Wear protective clothing when entering such areas.

8. *Avoid swimming in waters where snakes abound.* Most land species of poisonous snakes swim well, and may, under unusual circumstances, bite while in water. Sea snakes are not uncommon in the Indo-Pacific area, and while most species are docile some may bite when handled or disturbed.

9. *Avoid sleeping on the ground whenever possible.*

10. *Avoid walking close to rocky ledges.* Give snakes a wide passage, just in case.

11. *Avoid hiking alone in snake-infested areas.*

12. *Avoid horse-play involving live or dead snakes.* Snakes should not be handled carelessly. Teasing people with snakes may have unexpected and unfortunate results.

Specific Precautions

The following *DON'Ts* are suggested for those in snake country.

1. DON'T put your hands or feet in places you can not look, and

    DON'T put them in places without first looking.

2. DON'T turn or lift a rock or fallen tree with your hands. Move it with a stick, or with your foot if your ankle and leg are properly protected.

3. DON'T disturb snakes.

4. DON'T put your sleeping bag near rock piles or rubbish piles or near the entrance to a cave.

5. DON'T sit down without first looking around carefully.

6. DON'T gather firewood after dark.

7. DON'T step over a log if the other side is not visible. Step on it first.

8. DON'T enter snake-infested areas without adequate protective clothing.

9. DON'T handle freshly killed venomous snakes. Always carry them on a stick or in a bag if they must be returned to the command post.

10. DON'T crawl under a fence in high grass, or in an uncleared area.

11. DON'T go out of your way to kill a snake. Thousands of people are bitten by snakes each year merely because they try to kill them without knowing anything of their habits or habitats.

12. Finally, DON'T PANIC!

NOTES

Chapter III

HOW TO RECOGNIZE SNAKE VENOM POISONING
Symptoms and Signs

INTRODUCTION

In most parts of the world, bites by nonvenomous snakes occur far more frequently than bites by venomous snakes. Since the differentiation is often difficult, all victims of snakebite should be brought under the care of a physician as quickly as possible. Whenever feasible the offending snake should be killed and brought with the victim to the physician or person charged with the responsibility of identifying the reptile.

While it is not always possible to identify the snake responsible for the bite by the tooth or fang marks found on the victim's skin, in some cases these may be of considerable value in differentiating between bites by venomous and nonvenomous species. Bites by the vipers (Old World vipers, pit vipers of Asia, eastern Europe, and the rattlesnakes and related species of the Americas) usually result in one or two relatively large puncture wounds of varying depth, depending on the size of the snake, the force of its strike, and other factors. In most cases, additional tooth marks are not seen. Bites by the elapid snakes (cobras, mambas, tiger snake, taipan, coral snakes and related species) usually produce one or two small puncture wounds, although occasionally there may be one or two additional punctures. Sea snake bites are characterized by multiple (2 to 20) pinhead-sized puncture wounds. In some cases the teeth may be broken off and remain in the wound.

Proper identification of fang or tooth marks may be complicated in those cases where skin tears result from jerking an extremity away during the biting act. This is a particular problem in viper bites where long scratches or even lacerations are inflicted by the fangs. In bites by elapid snakes there may be superficial scratches from the snake's mandibular and palatine teeth. Thus, it can be seen that while fang or tooth patterns may be of assistance in determining the identity of an offending snake, they should not be depended upon as the deciding factor in establishing the diagnosis.

It should be noted that *one can be bitten by a venomous snake and not be poisoned.* In 3 to 40 per cent of the bites inflicted by venomous snakes, no signs or symptoms of poisoning develop. This may be due to the fact that the snake does not always eject venom or, if venom is ejected, that it does not enter the wound, as can sometimes happen in very superficial bites. This important fact should always be considered before specific treatment is started.

Venom Apparatus

The venom apparatus of a snake consists of a gland, a duct, and one or more fangs located on each side the the head (fig. 1). The size of these structures depends on the size and species of the snake. Each venom gland is invested in a connective tissue sheath which is invaded by the muscles that contract it during discharge of the venom. The innervation of these muscles is different from that controlling the biting mechanisms; thus, the snake can control the amount of venom it ejects. It can discharge venom from either fang, from both, or from neither. Snakes rarely eject the full contents of their glands.

Most rattlesnakes probably discharge between 25 and 75 percent of their venom when they bite a human. The true vipers discharge about the same, perhaps slightly less. There appears to be a greater variation in the amount an elapid may discharge. Many victims of elapid venom poisoning have minimal signs and symptoms; others show evidence of severe poisoning.

The fangs of the vipers are two elongated, canaliculated teeth of the maxillary bones. These bones can be rotated so that the fangs can be moved from their resting positions against the upper jaw, to their biting positions, approximately perpendicular to the upper jaw. These snakes have full control over their fangs, raising or lowering them at will as when striking, biting, or yawning. The two functional fangs are shed periodically and are replaced by the first reserve fangs. The fangs of the elapid snakes are two enlarged anterior maxillary teeth. These teeth are hollow and are fixed in an erect position.

Snake Venoms

The venom of most snakes is a complex mixture, chiefly proteins, many of which have enzymatic activity. Some of the effects of snake venoms are due to the nonenzymatic protein portions of the venom, while others are due to the enzymes and enzymatic combinations. The symptoms and signs of snake venom poisoning may be complicated by the release of several substances from the victim's own tissues. These autopharmacologic substances sometimes render diagnosis and treatment more difficult.

The arbitrary division of venoms into such groups as *neurotoxins, hemotoxins, and cardiotoxins*, while having some useful purpose in classification, has led to much misunderstanding and a number of errors in treatment. It has become increasingly apparent that these divisions are over-simplified and misleading. Neurotoxins can, and often do, have cardiotoxic or hemotoxic activity, or both; cardiotoxins may have neurotoxic or hemotoxic activity, or both; and hemotoxins may have the other activities. It is best to consider *all* snake venoms capable of producing several changes, sometimes concomitantly, in one

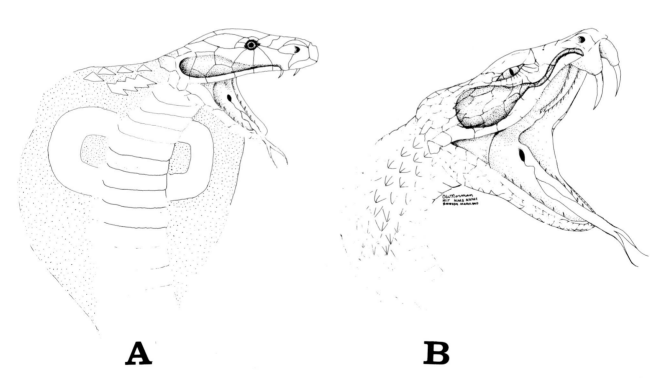

A   B

FIGURE 1.—Figures of fangs, venom ducts, and venom glands of: A. Cobra (Elapidae), and B. Viper (Viperidae). The fangs of elapid snakes are much shorter than those of vipers and do not rotate. In each case the venom glands lie outside the main jaw muscles toward the back of the head. The venom ducts lead from the glands to the bases of the hollow fangs.

TABLE 1.—YIELD AND LETHALITY OF VENOMS OF IMPORTANT POISONOUS SNAKES

Snake	Average length of adult (inches)	Approximate yield, dry venom (mg.)	Intraperitoneal LD$_{50}$ (mg./kg.)	Intravenous LD$_{50}$ (mg./kg.)
North America				
A. Rattlesnakes (*Crotalus*)				
Eastern diamondback (*C. adamanteus*)	33–65	370–720	1.89	1.68
Western diamondback (*C. atrox*)	30–65	175–325	3.71	4.20
Timber (*C. horridus horridus*)	32–54	95–150	2.91	2.63
Prairie (*C. viridis viridis*)	32–46	25–100	2.25	1.61
Great Basin (*C. v. lutosus*)	32–46	75–150	2.20	—
Southern Pacific (*C. v. helleri*)	30–48	75–160	1.60	1.29
Red diamond (*C. ruber ruber*)	30–52	125–400	6.69	3.70
Mojave (*C. scutulatus*)	22–40	50–90	0.23	0.21
Sidewinder (*C. cerastes*)	18–30	18–40	4.00	—
B. Moccasins (*Agkistrodon*)				
Cottonmouth (*A. piscivorus*)	30–50	90–148	5.11	4.00
Copperhead (*A. contortrix*)	24–36	40–72	10.50	10.92
Cantil (*A. bilineatus*)	30–42	50–95	—	2.40
C. Coral snakes (*Micrurus*)				
Eastern coral snake (*M. fulvius*)	16–28	2–6	0.97	—
Central and South America				
A. Rattlesnakes (*Crotalus*)				
Cascabel (*C. durissus terrificus*)	20–48	20–40	0.30	—
B. American lance-headed vipers (*Bothrops*)				
Barba amarilla (*B. atrox*)	46–80	70–160	3.80	4.27
C. Bushmaster (*Lachesis mutus*)	70–110	280–450	5.93	—
Asia				
A. Cobras (*Naja*)				
Asian cobra (*N. naja*)	45–65	170–325	0.40	0.40
B. Kraits (*Bungarus*)				
Indian krait (*B. caeruleus*)	36–48	8–20	—	0.09
C. Vipers (*Vipera*)				
Russell's viper (*V. russelii*)	40–50	130–250	—	0.08
D. Pit vipers (*Agkistrodon*)				
Malayan pit viper (*A. rhodostoma*)	25–35	40–60	—	6.20
Africa				
A. Vipers				
Puff adder (*Bitis arietans*)	30–48	130–200	3.68	—
Saw-scaled viper (*Echis carinatus*)	16–22	20–35	—	2.30
B. Mambas (*Dendroaspis*)				
Eastern green mamba (*D. angusticeps*)	50–72	60–95	—	0.45
Australia				
A. Tiger snake (*Notechis scutatus*)	30–56	30–70	0.04	—
Europe				
A. Vipers				
European viper (*Vipera berus*)	18–24	6–18	0.80	0.55
Indo-Pacific				
A. Sea snakes				
Beaked sea snake (*Enhydrina schistosa*)	30–48	7–20	—	0.01

or more of the organ systems of the body.

It is also apparent that quantitative and, perhaps, qualitative differences in the chemistry of venoms may occur at the species level and may, in fact, be evident in snakes of the same species taken from different geographic areas. Thus, differences in the symptoms and signs of poisoning may occur even when similar snakes are involved in a series of accidents.

In Table 1 are given the names of some of the more important venomous snakes of the world, their adult average lengths, the approximate amount of dried venom contained within their venom glands (adult specimens), and the intraperitoneal and intravenous LD$_{50}$ in mice, as expressed in milligrams of venom (on a dry weight basis) per kilogram of test animal body weight. The purpose of this table is to demonstrate the considerable differences that exist in the lethality of various snake venoms.

In general, the venoms of the vipers cause deleterious changes in the tissues both at the site of the bite and in its proximity, changes in the red blood cells, defects in coagulation, injury to the blood vessels; and, to a lesser extent, damage to the heart muscle, kidneys, and lungs. The venom of the tropical rattlesnake, *Crotalus durissus*, causes more severe changes in nerve conduction and neuromuscular transmission than do other crotalid venoms. The venoms of the elapid snakes cause lesser local tissue changes, but often cause serious alterations in sensory and motor function as well as cardiac and respiratory difficulties.

SYMPTOMS AND SIGNS

The symptoms, signs, and the gravity of snake venom poisoning are dependent upon a number of factors: the age and size of the victim, the nature, location, depth, and number of bites, the length of time the snake holds on, the extent of anger or fear that motivates the snake to strike, the amount of venom injected, the species and size of the snake involved, the condition of its fangs and venom glands, the victim's sensitivity to the venom, the pathogens present in the snake's mouth, and the degree and kind of first aid and subsequent medical care. It can be seen that

snakebites may vary in severity from trivial to extremely grave.

The findings given in tables 2, 3, and 4 are those observed in what may be termed typical, moderately severe cases of snake venom poisoning. While they are not complete, they do provide a ready reference of the more commonly observed sequelae of bites by venomous snakes.

Diagnosis of crotalid envenomation is dependent upon the presence of one or more fang marks, and immediate and usually progressive swelling, edema, and pain. In most cases swelling and edema are constant findings and are usually seen about the injured area within 10 minutes of the bite. In the absence of treatment, the swelling progresses rapidly and may involve the entire injured extremity within one hour. Generally, however, swelling and edema spread more slowly, and usually over a period of 8 to 36 hours. Swelling and edema are most marked following bites by the North American rattlesnakes (excluding the Mojave, massasaugas, and pigmy rattlesnakes) and the American lance-headed vipers (*Bothrops*). Swelling is slightly less marked following bites by the Malayan pit viper (*Agkistrodon rhodostoma*) and related species, the Asian lance-headed vipers (*Trimeresurus*), and the American moccasins (*Agkistrodon*). It is least acute following bites by the cascabel (*Crotalus durissus terrificus*).

In many cases, discoloration of the skin and ecchymosis appear in the area of the bite within several hours. The skin appears tense and shiny. Vesicles may form within 3 hours, and are generally present by the end of 24 hours. Hemorrhagic vesiculations and petechiae are common.

Pain immediately following the bite is a common complaint in most cases of crotalid poisoning. It is most severe following bites by the South American pit vipers (except for the cascabel, which is less severe); the eastern diamondback, western diamondback, and timber rattlesnakes of North America, and the Asian lance-headed vipers.

Weakness, sweating, faintness, and nausea are commonly reported. Regional lymph nodes may be enlarged, painful, and tender. A very common complaint following bites by some rattlesnakes, and one sometimes reported following other pit viper bites, is tingling or numbness over

TABLE 2.—SYMPTOMS AND SIGNS OF CROTALID BITES

Symptoms and Signs [1]	North American Rattlesnakes (*Crotalus*)	Central and South American Rattlesnakes (*Crotalus*)	North American Moccasins (*Agkistrodon*)	American Lance-headed Vipers (*Bothrops*)	Asian Lance-headed Vipers (*Trimeresurus*)	Malayan Pit Viper (*Agkistrodon*)
Swelling and edema	+++ [5]	+	++	+++	+++	+++
Pain	++	++	+	+++	++	+++
Discoloration of skin	+++	++	+	+++	+++	+++
Vesicles	+++		++	+++	++	+++
Ecchymosis	+++	+	++	+++	+++	+++
Superficial thrombosis	++		−	++	−	−
Necrosis	++		+	+++	+	+
Sloughing of tissue	++		−	+++	+	+
Weakness	++	+++	+	++	++	++
Thirst	++	++	+	+++	+	++
Nausea or vomiting or both	++	++	+	++	−	−
Diarrhea	+	++	−	++	†	−
Weak pulse and changes in rate	+++	+++	+	+++	++	++
Hypotension or shock	+++	+	+	+++	++	++
Sphering or destruction of red blood cells	+++	−	−	+++	−	−
Increased bleeding time	++	+	−	+++	+	+
Increased clotting time	+++	+	−	+++	+++	+++
Hemorrhage [2]	++		+	+++	++	+++
Anemia	++		−	+++	+	+
Blood platelet changes [3]	++		−	+++	+	+
Glycosuria	++		+	+++	−	−
Proteinuria	++	+	++	++	++	+
Tingling or numbness [4]	++	++	+++	++	+	−
Fasciculations	+	+++	−	+	−	−
Muscular weakness or paralysis	+	+++	−	+	−	−
Ptosis	+	+++	−	+	−	−
Blurring of vision	+	+++	−	+	−	−
Respiratory distress	++	+++	++	++	−	−
Swelling regional lymph nodes	++	+	+	++	+	−
Abnormal ECG	+	++	−	+	++	++
Coma	+	++	−	+	+	+

[1] In the more severe cases the intensity of the symptoms and signs may be markedly increased. In addition, there may be severe respiratory distress, cyanosis, muscle spasms, and secondary shock leading to death.

[2] Bleeding may be from the gastrointestinal, urinary, or respiratory tracts, from the gums, or it may be subcutaneous. In *Trimeresurus* bites the hemorrhage is usually confined to the locus of the wound. Bleeding from the gums is common following *Bothrops* envenomation.

[3] Platelets may be increased in mild poisonings and markedly decreased in severe cases.

[4] Often confined to the tongue and mouth, but may involve the scalp and distal parts of the toes and fingers as well as the injured part.

[5] (+ to +++) = Grading of severity of symptom, sign or finding, (−) = Of lesser significance or absent, ( ) = Information lacking.

TABLE 3.—SYMPTOMS AND SIGNS OF VIPERID BITES

Symptoms and Signs	Russell's Viper (*Vipera russelii*)	Saw-Scaled Viper (*Echis carinatus*)	Levantine Viper (*Vipera lebetina*) and related species	European Viper (*Vipera berus*)	Puff Adder (*Bitis arietans*)
Swelling and edema	+++	+++	++	++	++
Pain	+++	+++	++	++	+++
Discoloration of skin	+++	++	++	++	+++
Weakness	++	++	+	++	++
Nausea or vomiting or both	++	+	+++	++	++
Abdominal pain	++	+	+++	+	++
Diarrhea	++	+	+++	+	++
Thirst	++	+++	+	++	+
Chills or fever	++	++	++	-	+
Swelling regional lymph nodes	+	+	++	++	++
Facial edema	+	-	++	+	+
Dilatation of pupils	++	+	+	+	+
Weak pulse and changes in rate	++	+	++	+	+
Albuminuria	++	+	++	-	-
Proteinuria	++	++	++	-	-
Hypotension	++	++	++	+	++
Shock	++	++	++	+	++
Hemorrhage [1]	++	+++	++	+	++
Anemia	++	++	+	-	-
Vesicles	++	++	++	+	++
Ecchymosis	++	++	+	+	++
Necrosis	++	+	+	-	++
Decreased platelets	+	+	+	-	+
Prolonged clotting time	++	+++	++	-	+

[1] Usually limited to area of wound in puff adder and European viper bites. However, bleeding from the gums, intestine and urinary tract may occur, particularly in saw-scaled viper and Russell's viper envenomations.

TABLE 4.—SYMPTOMS AND SIGNS OF ELAPID BITES

Symptoms and Signs	Cobras (*Naja*)	Kraits (*Bungarus*)	Mambas (*Dendroaspis*)	Taipan (*Oxyuranus*)	Coral Snakes (*Micrurus*)
Pain	++	+	+	+	++
Localized edema	+	−	−	−	+
Drowsiness, weakness	+++	+++	++	+++	+++
Feeling of thickened tongue and throat, slurring of speech, difficulty in swallowing	+++	+++	+++	+++	++
Ptosis	++	+++	++	+++	++
Changes in respiration	++	+++	++	+++	++
Headache	++	++	++	++	++
Blurring of vision	++	++	++	+++	++
Weak pulse and changes in rate	++	++	++	+	++
Hypotension	++	++	++	+	+
Excessive salivation	+++	+++	+++	+	+++
Nausea and vomiting	+	++	+++	+++	+
Abdominal pain	+	+++	+++	+++	+
Pain in regional lymph nodes	++	++	+++	+++	+
Localized discoloration of skin	+	−	−	−	+
Localized vesicles	+	−	−	−	−
Localized necrosis	+	−	−	−	−
Muscle weakness, paresis or paralysis	++	+++	++	+++	+
Muscle fasciculations	+	+	+	+	+
Numbness of affected area	++	+++	++	+	+++
Shock	++	++	++	+	+
Convulsions	+	+	−	−	−

the tongue and mouth or scalp. Paresthesia about the wound is sometimes reported.

Viperid venom poisoning is characterized by burning pain of rapid onset, swelling and edema, and patchy skin discoloration and ecchymosis in the area of the bite. Extravasation of blood from the wound site is common in Russell's and saw-scaled viper envenomations. The failure of the blood to clot is a valuable diagnostic finding. Bleeding from the gums, and the intestinal and urinary tracts is common in severe Russell's and saw-scaled viper bites.

Cobra envenomation is characterized by pain usually within 10 minutes of the bite, and this is followed by localized swelling of slow onset, drowsiness, weakness, excessive salivation, and paresis of the facial muscles, lips, tongue, and larynx. The pulse is often weak, blood pressure is reduced, respirations are labored, and there may be generalized muscular weakness or paralysis. Ptosis, blurring of vision, and headache may be present. Contrary to popular opinion, necrosis is not an uncommon consequence of cobra venom poisoning. In bites by the kraits a similar clinical picture is usually seen, except that there is very little or no local swelling or severe pain. The systemic manifestations may often be more severe, and shock, marked respiratory depression and coma, may rapidly develop. Abdominal pain is often intense following poisoning by the kraits, mambas, and taipans. Envenomation by coral snakes may resemble krait venom poisoning. The bite is usually less painful, and there is occasionally a sensation of numbness about the wound. Chest pain, particularly on inspiration, is sometimes reported. Localized edema is minimal and necrosis is rare.

Mamba venom poisoning is characterized by weakness, nausea and vomiting, blurred vision, slurred speech, excessive salivation, headache, and abdominal pain. These findings are often followed by hypotension, respiratory distress, and shock.

Envenomation by most of the Australian-Papuan elapids produces drowsiness, visual disturbances, ptosis, nausea and vomiting, headache, abdominal pain, slurring of speech, respiratory distress, and generalized muscular weakness or paralysis. Hemoglobinuria may be found early in the course of the poisoning.

Sea snake venom poisoning is usually characterized by multiple pinhead-sized puncture wounds, little or no localized pain, oftentimes tenderness and some pain in the skeletal muscles and, in particular, the larger muscle masses and the neck. This pain is increased with motion. The tongue feels thick and its motion may be restricted. There may be paresthesia about the mouth. Sweating and thirst are common complaints, and the patient may complain of pain on swallowing. Trismus, extraocular weakness or paralysis, dilatation of the pupils, ptosis and generalized weakness may be present. Respiratory distress is common in severe cases. Myoglobinuria is diagnostic.

Little is known about the problem of envenomation by rear-fanged colubrid snakes. The African boomslang and bird snake are known to produce severe poisoning, which on rare occasions may be fatal. (These snakes are described on pp. 90–91.) Other species of colubrids, some with enlarged grooved fangs and some with solid teeth, are known to bite and may be venomous. The manifestations of poisoning by known venomous colubrids, such as the mangrove snake (*Boiga dendrophila*) of southeast Asia, the West Indian racers (*Alsophis*), the "culebra de cola corta" (*Tachymensis peruviana*) of western South America, the parrot snakes (*Leptophis*) of tropical America and several other species are local pain and swelling, sometimes accompanied by localized skin discoloration and ecchymosis; and in the more severe envenomations, increased swelling and edema which may involve the entire injured extremity, general malaise and fever. The acute period of the poisoning may persist for 4 to 7 days. It is important to differentiate envenomation by colubrids from that by the more dangerous elapids and vipers.

In summary, any snakebite associated with immediate (and sometimes intense) pain, and followed within several minutes by the appearance of swelling and subsequently edema is usually diagnostic of snake venom poisoning by a viper. Elapid envenomation, on the other hand, is not so easily diagnosed during the first 10 minutes following the bite. Pain, usually of minor intensity, may appear within the first 10 minutes, although in some cases it is not reported for 30 minutes or even longer. Swelling usually appears 2 or 3 hours following the bite and tends to be limited to the area of the wound. The first systemic sign of elapid venom poisoning is usually drowsiness. This is often apparent within 2 hours of the bite. Ptosis, blurring of vision, and difficulties in speech and swallowing may also appear within several hours of the bite. It can be seen how important it is in cobra, mamba, krait, taipan, tiger, and coral snake bites to determine the identity of the offending reptile as quickly as possible. A difference of 30 minutes to 1 hour in initiating treatment in elapid venom poisoning may make the difference between life and death.

REFERENCES

CAMPBELL, C. H. 1964. Venomous Snake Bite in Papua and Its Treatment with Tracheotomy, Artificial Respiration and Antivenene. Trans. R. Soc. Trop. Med. Hyg. 58 : 263-273.

CHRISTENSEN, P. A. 1955. South African Snake Venoms and Antivenoms. South African Institute Medical Research, Johannesburg, 142 p.

EFRATI, P. and REIF, L. 1953. Clinical and Pathological Observation on Sixty-five Cases of Viper Bite in Israel. Amer. J. Trop. Med. Hyg. 2 : 1085-1108.

GENNARO, J. F., Jr. 1963. Observations on the Treatment of Snakebite in North America, p. 427–446. In, H. L. Keegan and W. V. Macfarlane, Venomous and Poisonous Animals and Noxious Plants of the Pacific Region. Pergamon, Oxford.

HEATWOLE, H. and BANUCHI, I. B. 1966. Envenomation by the Colubrid Snake, *Alsophis portoricensis*. Herpetologica 22 : 132–134.

KAISER, E. and MICHL, M. 1958. Die Biochemie der Tierischen Gifte. F. Deuticke, Wien, 258 p.

KLAUBER, L. M. 1956. Rattlesnakes, Their Habits, Life Histories, and Influence on Mankind. University California Press, Berkeley. 2 vol.

MOLE, R. H. and EVERARD, A. 1947. Snakebite by *Echis carinata*. Quart. J. Med. 16 : 291–303.

REID, H. A. 1961. Myoglobinuria and Seasnake-bite Poisoning. Brit. Med. J. 1 : 1284–1289.

REID, H. A., THEAN, P. C., CHAN, K. E. and BAHARAM, A. R. 1963. Clinical Effects of Bites by Malayan Viper (*Ancistrodon rhodostoma*). Lancet 1 : 617–621.

RUSSELL, F. E. 1962. Snake Venom Poisoning, vol II, p. 197–210. In, G. M. Piersol, Cyclopedia of Medicine, Surgery and the Specialties. F. A. Davis, Philadelphia.

SAWAI, Y., MAKINO, M., TATENO, I., OKONOGI, T. and MITSUHASHI, S. 1962. Studies on the Improvement of Treatment of Habu Snake (*Trimeresurus flavoviridis*) Bite. 3. Clinical Analysis and Medical Treatment of Habu Snake Bite on the Amami Islands. Jap. J. Exp. Med. 32 : 17–138.

SCHENONE, H. and REYES, H. 1965. Animales ponzonosos de Chile. Bol. Chileno de Parasitol. 20 : 104–108.

WALKER, C. W. 1945. Notes on Adder-bite (England and Wales). Brit. Med. J. 2 : 13–14.

PLATE I
Representative American Pit Vipers (CROTALIDAE)

FIGURE 1. Rock Rattlesnake, *Crotalus lepidus*. Photo by Hal B. Harrison: National Audubon. (See pp. 36, 50)

FIGURE 2. Massasauga, *Sistrurus catenatus*. Photo by Charles Hackenbrock and Staten Island Zoo. (See p. 44)

FIGURE 3. Cascabel, *Crotalus durissus*. Photo by Roy Pinney and Staten Island Zoo. (See p. 68)

FIGURE 4. Cantil, *Agkistrodon bilineatus*. Photo by Hal B. Harrison: National Audubon. (See p. 54)

FIGURE 5. Broad-banded American Copperhead, *Agkistrodon contortrix* subspecies *laticinctus*. Photo by J. Markham. (See p. 39)

FIGURE 6. Cottonmouth, *Agkistrodon piscivorus*. Photo by J. Markham. (See p. 39)

PLATE II
Representatives of Some Poisonous Snake Families

FIGURE 1. European Viper, *Vipera berus* (VIPERIDAE). Photo by J. Markham. (See p. 74)

FIGURE 2. Urutu, *Bothrops alternatus* (CROTALIDAE). Photo by J. Markham. (See p. 65)

FIGURE 3. River Jack, *Bitis nasicornis* (VIPERIDAE). Photo by J. Markham. (See p. 102)

FIGURE 4. Puff Adder, *Bitis arietans* (VIPERIDAE). Photo by J. Markham. (See p. 101)

FIGURE 5. Eastern Coral Snake, *Micrurus fulvius* (ELAPIDAE). Photo by Allan D. Cruickshank: National Audubon. (See p. 38)

PLATE III
Representatives of Some Poisonous Snake Families

FIGURE 1. Black Mamba, *Dendroaspis polylepis* (ELAPIDAE). Photo by J. Markham. (See p. 93)

FIGURE 2. Western Diamondback Rattlesnake, *Crotalus atrox* (CROTALIDAE). Navy photo, courtesy U.S. National Zoological Park. (See p. 40)

FIGURE 3. Prairie Rattlesnake, *Crotalus v. viridis* (CROTALIDAE). Navy photo, courtesy U.S. National Zoological Park. (See p. 42)

FIGURE 4. Timber Rattlesnake, *Crotalus horridus* (CROTALIDAE). Navy photo, courtesy U.S. National Zoological Park. (See p. 43)

FIGURE 5. Cottonmouth, *Agkistrodon piscivorus* (CROTALIDAE). Navy photo, courtesy U.S. National Zoological Park. (See p. 39)

FIGURE 6. American Copperhead, *Agkistrodon contortrix* (CROTALIDAE). Navy photo, courtesy U.S. National Zoological Park. (See p. 39)

PLATE IV
Representative Pit Vipers (CROTALIDAE)

FIGURE 1. American Copperhead, *Agkistrodon contortrix*, southeastern U.S. Navy photo, courtesy U.S. National Zoological Park. (See p. 39)

FIGURE 2. Chinese Green Tree Viper, *Trimeresurus stejnegeri*. Navy photo, courtesy U.S. National Zoological Park. (See p. 129)

FIGURE 3. Green Tree Viper, *Trimeresurus* sp. Navy photo, courtesy U.S. National Zoological Park. (See p. 130)

FIGURE 4. Wagler's Pit Viper, *Trimeresurus wagleri*. Navy photo, courtesy U.S. National Zoological Park. (See p. 138)

FIGURE 5. Okinawa Habu, *Trimeresurus flavoviridis*. Navy photo, courtesy U.S. National Zoological Park. (See p. 137)

FIGURE 6. Sakishima Habu, *Trimeresurus elegans*. Navy photo, courtesy U.S. National Zoological Park. (See p. 137)

PLATE V

Some Poisonous Snakes of Asia

FIGURE 1. Sharp-nosed Pit Viper, *Agkistrodon acutus* (CROTALIDAE). Navy photo, courtesy U.S. National Zoological Park. (See p. 136)

FIGURE 2. Many-banded Krait, *Bungarus multicinctus* (ELAPIDAE). Navy photo, courtesy U.S. National Zoological Park. (See p. 120)

FIGURE 3. Chinese Mountain Viper, *Trimeresurus monticola* (CROTALIDAE). From a painting. (See p. 138)

FIGURE 4. Chinese Habu, *Trimeresurus mucrosquamatus* (CROTALIDAE). From a painting. (See p. 137)

FIGURE 5. Japanese Mamushi, *Agkistrodon halys blomhoffii* (CROTALIDAE). From a painting. (See p. 136)

FIGURE 6. MacClelland's Coral Snake, *Calliophis macclellandii* (ELAPIDAE). From a painting. (See p. 122)

PLATE VI

Some Poisonous Snakes of Asia

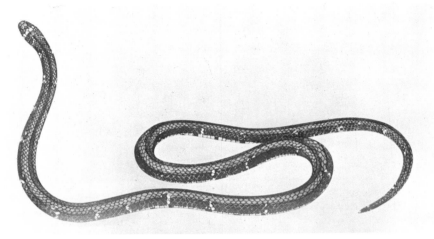

FIGURE 1. Asian Coral Snake, *Calliophis sauteri*
(ELAPIDAE). From a painting. (See p. 122)

FIGURE 2. Chinese Cobra, *Naja naja atra* (ELAPI-
DAE). Navy photo, courtesy U.S. National
Zoological Park. (See p. 125)

FIGURE 3. Russell's Viper, *Vipera russelii* (VI-
PERIDAE). Navy photo. (See p. 127)

FIGURE 4. Annulated Sea Snake, *Hydrophis
cyanocinctus* (HYDROPHIDAE). Photo by
Sherman A. Minton. (See p. 165)

FIGURE 5. Indian Krait, *Bungarus caeruleus*
(ELAPIDAE). Photo by Sherman A. Minton.
(See p. 120)

PLATE VII. Some Poisonous Snakes of Africa

(From Pitman's *Snakes of Uganda*)

FIGURE 1. Boomslang, *Dispholidus typus* (COLUBRIDAE). Brown coloration. Note lack of distinct pattern. (See p. 90)

FIGURE 2. Gold's Tree Cobra, *Pseudohaje goldii* (ELAPIDAE). The very glossy appearance of scales is not well shown here. (See p. 97)

FIGURE 3. Forest Cobra. *Naja melanoleuca* (ELAPIDAE). Juvenile coloration. (See p. 96)

FIGURE 4. Boomslang, *Dispholidus typus* (COLUBRIDAE). Juvenile coloration. (See p. 90)

FIGURE 5. Green Bush Viper, *Atheris squamigera* (VIPERIDAE). (See p. 99)

FIGURE 6. Sedge Viper, *Atheris nitschei* (VIPERIDAE). (See p. 98)

FIGURE 7. Black Mole Viper, *Atractaspis irregularis* (VIPERIDAE). (See p. 99)

FIGURE 8. Günther's Mole Viper, *Atractaspis aterrima* (VIPERIDAE). (See p. 99)

FIGURE 9. Banded Water Cobra, *Boulengerina annulata* (ELAPIDAE). This eastern subspecies is unicolor over most of body. (See p. 92)

FIGURE 10. Jameson's Mamba, *Dendroaspis jamesoni* (ELAPIDAE). Widespread arboreal mamba with a usually dark, often black, tail. (See p. 93)

FIGURE 11. Black Mamba, *Dendroaspis polylepis* (ELAPIDAE). Ground color varies between gray and brown. Scales never show gloss that they do in some cobras. (See p. 93)

PLATE VIII. Some Poisonous Snakes of Africa

(From Pitman's Snakes of Uganda)

FIGURE 1. Bird Snake, *Thelotornis kirtlandii* (COLUBRIDAE). Note narrow pointed head, oblique dorsals, and lichen-like color pattern. (See p. 91)

FIGURE 2. Boomslang, *Dispholidus typus* (COLUBRIDAE). Green coloration. Note distinctly oblique dorsals and absence of distinct color pattern. (See p. 90)

FIGURE 3. African Garter Snake, *Elapsoidea sundevallii* (ELAPIDAE). See p. 94

FIGURE 4. Rhombic Night Adder, *Causus rhombeatus* (VIPERIDAE). Chevron-shaped head marking is distinctive. (See p. 102)

FIGURE 5. Green Night Adder, *Causus resimus* (VIPERIDAE). Oblique dorsals are characteristic of the genus. (See p. 102)

FIGURE 6. Lichtenstein's Night Adder, *Causus lichtensteinii* (VIPERIDAE). (See p. 102)

FIGURE 7. Egyptian Cobra, *Naja haje* (ELAPIDAE). Distinguished from other cobras by a row of subocular scales. (See pp. 80, 95)

FIGURE 8. Forest Cobra, *Naja melanoleuca* (ELAPIDAE). Adult coloration. Note highly polished appearance of scales. (See p. 96)

FIGURE 9. Spitting Cobra, *Naja nigricollis* (ELAPIDAE). Highly variable coloration; scales not as glossy as those of forest cobra. (See p. 96)

Chapter IV

FIRST AID

INTRODUCTION

Poisoning from snake venom is a medical emergency which requires immediate attention and the exercise of considerable judgment. Delayed or inadequate treatment for venomous snakebite may have tragic consequences. On the other hand, failure to differentiate between bites of venomous and nonvenomous snakes may lead to use of measures which bring not only discomfort to the individual but may produce deleterious results. It is essential that the one responsible for treatment establish whether or not envenomaiton has occurred before treatment is started. As was pointed out in Chapter III, a venomous snake may bite and not inject venom. Also, some persons bitten by nonvenomous snakes become excited and even hysterical. These emotions may give rise to disorientation, faintness, dizziness, rapid respiration or hyperventilation, rapid pulse, and even primary shock—all symptoms and signs which may occur following envenomation. The hospital corpsman should keep this syndrome in mind when called upon to treat a person bitten by an unidentified snake.

Most cases of snake venom poisoning in Navy and Marine Corps personnel have occurred in the presence of other service personnel. In most instances hospital corpsmen have been able to render the necessary initial first aid measures. However, in an occasional case no medically trained person may be available or the victim may be alone. As the success or failure of treatment may depend on *when* first aid is started, this chapter has been prepared to acquaint all Navy and Marine Corps personnel with the problem of snake venom poisoning and the first aid measures that need to be carried out in the event that poisoning occurs distant from a hospital, doctor, or medical corpsman.

GENERAL CONSIDERATIONS

Treatment, to be effective, must be instituted immediately following the bite and must include measures:
1. To retard absorption of the venom;
2. To remove as much venom as possible from the wound;
3. To neutralize the venom;
4. To prevent or reduce the effects of the venom; and
5. To prevent complications, including secondary infection.

If a victim of snakebite finds himself alone, a number of factors must be considered. He will need to determine whether or not the snake is venomous and, if venomous, the severity of the poisoning. He will need to consider how long it may be before help will reach him. Or, perhaps, he will need to weigh the advantages of walking to the nearest friendly troop facility, hospital, or town. If he decides to move he will need to determine how fast and for how long a period he may walk. These and other variables make it difficult to give consistent advice on what to do under such circumstances. Each case must be considered separately. The victim should give careful thought to all of these matters before making a decision as to the wisest procedure to follow. He should remember not to panic and not to overly exert himself. He should make every effort possible to obtain assistance, without jeopardizing his mission.

With these things in mind the following considerations should be followed, in so far as possible. The outcome of each case is dependent, among other things, upon the kind of treatment used and the speed with which it is initiated. Above all, the victim should remember to *keep cool*, and consider each move thoroughly.

STEP ONE

Apply a Constriction-band or a Tourniquet.—In cases of envenomation by most crotalids, a constriction-band should be placed above the first joint proximal, or 2 to 4 inches proximal to the bite, whichever is appropriate (see fig. 2). It should be applied tight enough to occlude the superficial venous and lymphatic return but not arterial flow. It should be released for 90 seconds every 10 minutes. The constriction-band can be moved in advance of the progressive swelling. It should be removed as soon as antivenin has been started. In no case of viper venom poisoning should a constriction-band be used for more than 4 hours. It is probably of little value if applied later than 30 minutes following the bite.

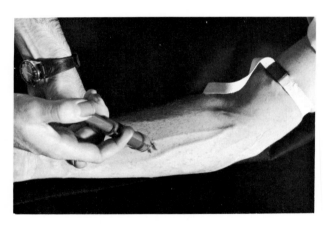

FIGURE 2.—Correct placement of a tourniquet. It should be tight enough to impede the flow of lymph and blood in the superficial vessels, but not that of blood in the deeper vessels.

Following envenomation by elapids, constriction-bands or tourniquets are of questionable value. However, in cases of severe envenomation by cobras, kraits, mambas, tiger snakes, death adders or taipans, a tight tourniquet should be applied immediately proximal to the bite and left in place until antivenin is given. It should be released for 90 seconds every 10 minutes, and should not be used for more than 8 hours.

STEP TWO

Capture the Snake and Kill It.—Most snakes will remain in the immediate area of the accident and can be found without too much difficulty. If several persons are present, send one or two in search of the snake while the others are administering first aid to the victim. Exercise extreme caution in hunting for the offending snake. *A reptile that has bitten once is just as likely to bite again as not.* The snake can be killed by a sharp blow on the neck. (An undamaged head is a great aid to identification). *Do not* handle the snake. If it cannot be positively identified at the scene of the accident, carry it on a stick or in a cloth bag to the command post or hospital.

STEP THREE

Lie Down.—Remain at rest until the offending snake has been identified (see Chapters VI, VII, and VIII). If the snake is nonvenomous, clean and dress the wound and proceed with your mission. Report to a medical officer as soon as possible.

If the snake is identified as venomous, or if its identity cannot be determined, begin treatment as outlined below:

STEP FOUR

Unidentified Snake.—Immobilize the injured part (see below) then turn to page 16 for instructions.

Identified Venomous Snake.—Immobilize the injured part. This can be done by splinting as for a broken arm or leg. The immobilized part should then be kept below the level of the heart, but not in a completely dependent position. If the wound is on the body, keep the victim in a sitting or lying position, depending on the location of the bite. The patient should always be kept warm. He should *not* be allowed to walk. He should *not* be given alcohol. He may, however, be given water, coffee, or tea. Any manifestations of fear or excitement should be alleviated by reassurance.

STEP FIVE

Make Incision and Apply Suction.—Incision and suction are of definite value when applied immediately following bites by vipers, particularly pit vipers of North America. They are of lesser value following bites by the South American vipers and Asiatic vipers, and probably of little value following envenomation by elapids and sea snakes (see discussion of first aid measures, p. 17).

In viper bites, excluding those by small European vipers and small North American copperheads, make cross-shaped or longitudinal incisions ⅛ to ¼ inches long through the fang marks (fig. 3), except in those cases where there is an abnormal amount of bleeding. The incisions should be made as deep as the fang penetration. The direction of the animal's strike and the curvature of the fang should be borne in mind when determining the plane of incision. Suction should then be applied and continued for the first hour following the bite. To be effective, suction must be applied within the first few minutes following the biting. It is of little value if delayed for 30 minutes or more. Oral suction should not be used if other means of suction are available. Multiple incisions over the involved extremity or in advance of progressive edema are not advised.

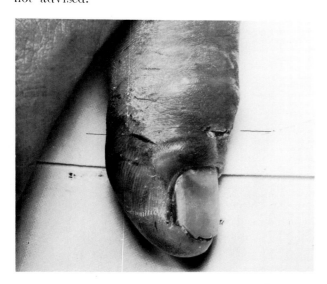

FIGURE 3.—Incised fang marks of a viper. Note how small the incisions through the wounds need to be. Photo by Findlay E. Russell.

STEP SIX

Administer Antivenin.—It is recommended that medical corpsmen, in the absence of a physician

and after suitable training, be given permission to conduct sensitivity tests and to administer appropriate antivenin to victims of snake venom poisoning. This might be done in those cases where severe signs and symptoms develop early in the course of the illness, or where 4 hours or more following viper venom poisoning or 2 hours or more following elapid venom poisoning can be expected to elapse before professional care will be available.

In such cases, following appropriate skin or eye tests (see Sensitivity Tests, page 16), the antivenin should be given intramuscularly at a site distant from the wound. *Antivenin should never be injected into a finger or toe*, and it should be administered intravenously only by qualified personnel. As the amount of antivenin available in the field is limited, one unit (vial or package) will probably be all that is available for a corpsman to give. The earlier this is injected, the better the results. However, several units may be needed for full neutralization of the venom.

NOTE: If the victim is in shock the antivenin will be absorbed slowly from an intramuscular site.

No Antivenin Available.—If antivenin is not available or if no qualified person is present to administer it, then proceed with STEP SEVEN.

STEP SEVEN

Transport Victim to Doctor, Aid Station or Hospital.—This should be done by litter, if at all possible; if not, try to provide some other means of transportation. Do not let the victim walk if this can be avoided. Keep the victim warm, and the bitten part in a dependent position.

STEP EIGHT

Institute Supportive Measures.—Should any of the following sequelae to the bite develop during evacuation of the victim, consider these measures:
Shock:

1. Place victim in recumbent or shock position (lying down on his back, head slightly lower than his feet).

2. Maintain an adequate airway.

3. Keep victim comfortably warm.

4. Control any severe pain. This can usually be done with salicylates or codeine. *Do not* give

morphine to an unconscious victim or one in respiratory distress.

5. Allay apprehension by reassuring words and actions.

6. Replace and maintain adequate blood volume with saline, plasma, plasma expanders, or whole blood. (Items 6, 7, and 8 are recommended for use by medical officers or paramedical personnel with appropriate qualifications).

7. Give vasopressor drugs if condition warrants.

8. Give oxygen.

Respiratory Distress:

1. Clear airway.

2. Apply artificial respiration. As long as the patient's heart continues to beat, he has a chance to recover, and this may occur even after many hours of artificial respiration. Mouth-to-mouth breathing in rhythm is the method of choice in all cases of respiratory failure. However, when it cannot be applied, the rhythmic push-pull methods are generally effective. If a mechanical resuscitator is available, it also may be used by anyone qualified in its operation.

3. Respiratory stimulants are limited to use by a medical officer.

Vomiting:

Vomiting frequently occurs following certain types of snake venom poisoning. Precautions should be taken to see that the patient does not aspirate vomitus. Place him in a prone position, head slightly lowered and turned to one side.

Excessive Salivation:

Place head in a position to permit adequate drainage of saliva as described under Vomiting, above. Keep airway clear. Atropine or parasympatholytic drugs may be administered only by a medical officer.

Convulsions:

No treatment should be given during the attack except that which will prevent the patient from injuring himself.

STEP NINE

Disposition of Patient.—At the aid station or hospital, inform the doctor of the identity of the snake involved (if known); or, turn the dead, unidentified snake over to the doctor. Give approximate time between bite and arrival and point out any constriction-band or tourniquet left in place. Give details on any antivenin or drugs given the patient. Report all unusual signs and symptoms.

BITES BY UNIDENTIFIED SNAKES

Every attempt should be made to capture and kill, or at least identify, the offending snake. As a rule, snakes remain in the vicinity of the accident. A knowledge of the habits and habitats of the snakes peculiar to the area (see Chapters VI, VII, and VIII) will assist in locating and identifying the snake. *But,* when the bite occurs at night, capture of the snake may not be possible, and management and treatment of the victim will depend upon the clinical signs and symptoms.

First, have the *victim lie down and remain at complete rest. Immobilize the bitten part and keep it in a dependent position.* In this position the onset of pain, if it occurs, will be more rapid, thus assisting in an early diagnosis. *Do not apply a tourniquet or incise the wound.* These measures may produce effects which could make diagnosis more difficult.

If no pain, swelling, edema, drowsiness, paresthesia, weakness, or paresis of the muscles of the face and throat appear within 30 minutes, the bite was probably inflicted by a nonvenomous snake. However, if at all possible, the victim should remain at rest and be observed for an additional 2 hours.

If symptoms or signs of venom poisoning develop during the observation period, the measures previously described under General Considerations, above, must be considered. The success of these measures will depend upon the time that has transpired between the bite, and their initiation. In those cases where the first aid measures have been deferred, the need for early administration of antivenin becomes urgent. In such cases, greater consideration should be given to the intravenous use of antivenin, obviously following the necessary sensitivity tests.

SENSITIVITY TESTS

A sensitivity test for horse serum must be carried out on all victims of snake venom poisoning before horse serum antivenin is administered. Directions for these tests will usually be found in the package containing the antivenin. In the absence of specific instructions follow these steps:

1. Inject 0.10 ml. of a 1:10 dilution of the horse serum or antivenin intracutaneously on the inner surface of the forearm. Use the specific hypodermic needle provided for the test. If one is not provided, use a short 27-gauge needle. If the test is done correctly, a wheal will be raised at the site of the injection. The wheal is white at first but if the test is positive the area about the point of injection will become red within 10 to 15 minutes. If any local or systemic allergic manifestations develop within 20 minutes of the test, *do not* give antivenin. Leave this decision to the medical officer.

If the victim develops a severe reaction to the test (restlessness, flushing, sneezing, urticaria, swelling of the eyelids and lips, respiratory distress or cyanosis), inject 0.3 to 0.5 ml. of 1:1,000 adrenalin subcutaneously, and observe the victim closely. Be prepared to administer artificial respiration. A cardiac stimulant may also be needed if shock develops.

2. An alternative to the skin test is the eye test. One or two drops of a 1:10 solution of the horse serum or antivenin are placed on the conjunctiva of one eye. If the test is positive, redness of the conjunctiva will develop within a few minutes. If the reaction is very severe, it should be controlled by depositing a drop or two of 1:1,000 adrenalin directly on the conjunctiva.

3. If a serum sensitivity test is positive, desensitization should be carried out before administering antivenin. This should be done *only* by a doctor. Do not attempt to desensitize a victim unless the appropriate facilities and drugs are available.

DISCUSSION OF FIRST AID MEASURES

It is not a purpose of this manual to discuss or evaluate all of the first aid treatments that have been suggested or advised for snake venom poisoning. This has been done in the medical literature. The reader is referred to the references at the end of this chapter for a more thorough consideration of this subject. It should be noted here, however, that there is no single thera-peutic standard of procedure for all cases of snake venom poisoning. Rest, immobilization of the injured part, and reassurance are indicated in every case, and in themselves are valuable therapeutic measures, but beyond these, few measures can be recommended for all cases of snakebite. In the following sections some consideration will be given to several of the more commonly employed first aid measures.

CONSTRICTION-BAND AND TOURNIQUET

Constriction-bands and tourniquets have long been used in the treatment of snakebite. The rationale for their use is quite simple, that is, to retard the absorption and spread of the venom. Studies with North American rattlesnake venom labeled with radioactive iodine (I^{131}) show that the spread of certain fractions of the venom can be retarded by pressure on the superficial lymphatic channels proximal to the deposition of the toxin. It appears that the greater portion of rattlesnake venom is absorbed directly into lymphatic structures. These studies support the clinical findings that in cases of North American rattlesnake bites, a constriction-band, applied early and effectively, can retard the spread of the toxin and thus decrease the area of localized necrosis. There is also some evidence to indicate that the constriction-band retards the development of systemic signs and symptoms.

The use of a constriction-band or a tourniquet in cases where deep envenomation has occurred would appear to be of limited or no value, and indeed some clinical reports support this contention. On the other hand, the incorrect application of constriction-bands and tourniquets, particularly in Southeast Asia, makes it difficult to evaluate these measures solely from clinical experiences. It might be concluded that a properly applied constriction-band is of definite value in poisoning by all North American crotalids and many of the small vipers from throughout the world. It is probably of lesser or no value following bites by the large vipers outside North America. In spite of these findings and opinions, it seems advisable, in view of no substantial contra-indication, to recommend the use of a constriction-band in all cases of viper venom poisoning during the period when suction is being carried out.

The venoms of the elapids are considerably different in their chemical structure from those of the vipers, and the structural variations within the venoms of the family Elapidae are definitely more complex than those within the families Crotalidae and Viperidae. Present knowledge indicates that, in general, elapid venoms are absorbed in greater quantities through the blood vessels than through lymphatic vessels.

Both experimental and clinical studies indicate that a constriction-band is of questionable value following envenomation by an elapid. The value of a tight tourniquet is not so easily decided. The rationale for using a tight tourniquet to occlude both superficial and deep blood vessels is easily understood. However, it must be admitted that adequate supportive evidence is still lacking. Nevertheless, it seems best to advise placing a tight tourniquet proximal to wounds inflicted by large cobras, kraits, mambas, tiger snakes, death adders, and taipans. Tourniquets should be left in place only until antivenin is injected. Under no circumstances should they be used for more than 8 hours, and never without the usual precautions associated with the use of a tourniquet.

INCISION AND SUCTION

Few problems in the first aid treatment of snake venom poisoning have elicited as much controversy as incision and suction. Recent experimental studies have shown that in the case of *Crotalus* envenomation, incision and suction at the fang puncture wounds instituted within several minutes of the bite, and suction continued for no less than 30 minutes, can remove a measurable portion of the venom. The exudate obtained from such incised wounds has been found to produce the typical fall in systemic arterial blood pressure, the increase in systemic venous and cisternal pressures, the changes in cardiac and respiratory rates, and the alterations in the electrocardiogram and electroencephalogram observed following injection of crude venom. The exudate has also been found to be lethal to mice in doses approaching that of the crude venom. Studies with I^{131}-labeled *Crotalus* venom indicate that the toxin can be removed from properly incised wounds by suction. These various experimental studies strongly support the clinical impressions of those physicians who have treated a sufficient number of rattlesnake bites to be in a position to evaluate this first aid measure critically.

Contrary to some opinion, few if any blood vessels, tendons, or other vital structures have been injured by properly executed cuts through fang marks of North American rattlesnakes. There is no foundation for the condemnation of this procedure on the basis that vital structures have been damaged during the execution of cuts. There is also no support for the contention that such trivial incisions will produce neural and glandular activities which, in turn, increase the lethal effect of the venom.

According to some clinicians, incision and suction through the fang wounds have not been found to be effective following the bites of vipers in Asia, Africa and parts of the Middle East. While they are advised and used by some physicians in these areas, others do not recommend their use. Adequately controlled studies on the depth of fang penetration by the Old World vipers have not been done, but clinical evidence would seem to indicate that these snakes bite deeper than their North American cousins. If this is true then incision and suction would be less effective than in North American crotalid bites. Where intramuscular envenomation occurs, incision and suction are of no value and are not recommended.

The time of instigation and the manner in which incisions have been made following bites by vipers in Asia and Africa have been so inconsistent that it is quite impossible to determine, solely on the basis of clinical reports, whether or not these measures are useful as first-aid measures in poisonings by the Old World vipers. It seems best at this time to advise incision and suction in most cases of viper venom poisoning. In no case, however, should incisions be made deeper than the subcutaneous tissues, and in those cases where it is obvious that the fangs have penetrated muscle tissue, no incisions should be made.

Incision and suction through the fang marks produced by the elapid snakes have not been found useful. This may be because elapid venoms are absorbed more directly into the blood stream than into lymphatic channels. It is not possible from the clinical reports on elapid bites to determine whether or not the measures are useful, since the time of making the cuts and the duration of the suction are seldom recorded. Also, observations in Asia and Africa indicate that these procedures are seldom carried out in what one might assume to be an effective manner. It would seem best to avoid using incision and suction for elapid venom poisoning until current experimental work on this problem has been completed, or until a critical evaluation of clinical cases has been made.

EXCISION

Excision of the bite area is a rather heroic measure which might be of value in some envenomations if it could be carried out within 2 or 3 minutes following the bite. It is a procedure carrying considerable risk. It might be considered in those cases where envenomation by a large krait, mamba, taipan, death adder, or tiger snake has occurred, and where the victim is alone and isolated, and likely to remain so for 6 or more hours. Under such conditions it might be wise to excise the wound or amputate the toe or finger. This has been done by some courageous persons.

OTHER MEASURES

According to Russell and Scharffenburg, some 217 "cures" for snake venom poisoning have been described in the literature. Some of the suggested first aid measures are: injecting potassium permanganate, ammonia, vinegar or oil into the wound; wrapping the liver of the offending snake or of a freshly-killed chicken over the wound; setting fire to the wound after applying gasoline; eating various plants or raw meat; applying mud packs to the wound; soaking the injured part in excrement; washing the wound with plant juices; drinking whiskey; taking antihistaminics, *et cetera*. These and the other so-called cures are little more than historical curiosities. Whatever the source, they are hazardous: first, because they often involve dangerous methods; second, because they delay the use of effective therapeutic procedures. They should not be used.

Snake venom poisoning is an accident highly variable in the gravity of its results. It is one in which the most fantastic remedy may gain its reputation among credulous people by having

"cured" a bite that required no treatment whatever. Avoid using any first aid measure that has not been evaluated; remember, most of the "cures" you will hear about have been evaluated and found to be useless.

REFERENCES

BUCKLEY, E. and PORGES, N. 1956. Venoms. Amer. Assoc. Advancement Sci., Washington, D.C. 467 p.

GENNARO, J. F., Jr. 1963. Observations on the Treatment of Snakebite in North America, p. 427–446. In, H. L. Keegan and W. V. Macfarlane, Venomous and Poisonous Animals and Noxious Plants of the Pacific Region. Pergamon, Oxford.

KAISER, E. and MICHL, M. 1958. Die Biochemie der Tierischen Gifte. F. Deuticke, Wien, 258 p.

KEEGAN, H. L. and MACFARLANE, W. V. (editors) 1963). Venomous and Poisonous Animals and Noxious Plants of the Pacific Region. Pergamon, Oxford. 456 p.

MERRIAM, T. W., Jr. 1961. Current Concepts in the Management of Snakebite. Mil. Med. 126:526–531.

NATIONAL ACADEMY OF SCIENCES, NATIONAL RESEARCH COUNCIL, COMMITTEE ON SNAKEBITE THERAPY. 1963. Interim Statement on Snakebite Therapy. Toxicon 1:81–87.

PHISALIX, M. 1922. Animaux Venimeux et Venins. Masson, Paris, 2 vol.

REID, H. A., THEAN, P. C., CHAN, D. E. and BAHARAM, A. R. 1963. Clinical Effects of bites by Malayan viper (*Ancistrodon rhodostoma*). Lancet 1:617–621.

RUSSELL, F. E. 1962. Snake Venom Poisoning, vol II, p. 197–210. In, G. M. Piersol, Cyclopedia of Medicine, Surgery and the Specialties. F. A. Davis, Philadelphia.

RUSSELL, F. E. and SCHARFFENBERG, R. S. 1964. Snake Venoms and Venomous Snakes. Bibliographic Associates, West Covina, California. 220 p.

Chapter V

MEDICAL TREATMENT

GENERAL CONSIDERATIONS

On arrival at the aid station or hospital, an immediate evaluation must be made of the patient's general condition. Snake venom poisoning is always a medical emergency requiring immediate attention. A delay in instituting medical treatment can lead to far more tragic consequences than one following an ordinary traumatic injury. In most cases, first aid measures will already have been instituted by the corpsman. The physician will need to evaluate these measures before determining the course of subsequent treatment. None of the first aid measures should be regarded as substitutes for antivenin, antibiotic and antitetanus agents; nor should they be instituted at the possible expense of delaying administration of the antivenin. Needless to say, the physician will have to establish in his own mind whether or not the patient has been poisoned and, if so, to determine which therapeutic measures he can use most effectively.

If the patient arrives at the medical installation one hour or more following the bite and no first aid measures have been initiated, the physician should put him to bed, immobilize the affected part, clean the wounds thoroughly, and proceed with the measures outlined below. Incision and suction, excision, *et cetera*, are of no value after such a delay and should not be attempted.

Admission Procedures

A routine history and physical examination should be done. The identity of the offending snake, its size, the time of the bite, and the details of all first aid measures employed, including the time lapse for each, should be recorded. Inquiry should be made concerning previous bites, allergies, and whether or not the patient has previously been exposed to horse serum. If a skin test has already been done this should be checked. Blood should be drawn for typing, cross-matching, blood clotting, and clot retraction studies. A complete blood count, hematocrit, and urine analysis are essential. Determinations of the sedimentation rate, prothrombin time, carbon dioxide combining power, urea nitrogen, sodium, potassium and chloride are advised, if within the limits of personnel, time, and equipment. In severe poisonings, an electrocardiogram and a blood platelet count should be done. Serum bilirubin, red cell fragility tests, and renal function tests should be done if the condition warrants. Studies of the hematocrit, complete blood counts, and hemoglobin concentration should be carried out several times a day. Urinalyses should be done with particular attention being given to the presence or absence of red cells. This is especially important in all cases of viper venom poisoning.

In all patients, regardless of the snake involved, pulse, blood pressure, and respirations should be checked periodically. When available, central venous pressure monitoring devices may be used in order to determine need for and to evaluate response to anti-shock therapy. Facilities and drugs for shock must be readily available, and a tracheostomy set and positive pressure breathing apparatus should be held in readiness. A measurement of the circumference of the affected part 4 inches above the bite, and at an additional point proximal to the wound, should be recorded.

The course of snake venom poisoning is sometimes unpredictable, and patients showing steady recovery may on occasions take a turn for the worse. Continued close observation by physicians and nurses is essential during the entire hospitalization period.

SPECIFIC THERAPY

Antivenin

The early administration of antivenin, particularly following a severe envenomation, cannot be overemphasized. A few minutes may mean the difference between life and death. The choice of antivenin, the route of injection and the amount to be given will depend upon a number of different factors (see below). In most cases, the more species or genus specific the antivenin, the more effective it will be. However, at the present time there is a great deal of variation in the effectiveness of the commercially available antivenins; some polyvalent types appear to be more useful than some which are genus specific. Unfortunately, there is no standardized process for the production of antivenin, and indeed there is no conformity in testing methods. Thus, the physician will need to depend on the specific information supplied with the antivenin, or upon more detailed data provided by a medical facility in the area. Ampoules of antivenin usually have an "expiration date" indicated. Though this is the limit of the producer's period of potency, the antivenin does not suddenly become ineffective on that date. Some producers have indicated that the effectiveness of the antivenin is not greatly impaired until it has become cloudy or milky in appearance. A list of the available antivenins is provided on pages 169–179.

Certain principles can, however, guide the physician in his choice of an antivenin. In general, the lyophilized preparation is to be preferred to the nonlyophilized one; and antivenins prepared by fractionation with ammonium sulfate, or some similar process for removing the low antibody containing fractions, are usually superior to those in which the whole serum is packaged. Almost all antivenins currently available are prepared in horses, but within the next few years some antivenins will be prepared from sera of other animals.

The amount of antivenin required to neutralize the effects of a venom will depend upon a number of different factors. However, some general instructions can be given. Following appropriate skin or eye tests, in cases of minimal envenomation, 1 or 2 units (vials, tubes or packages) will usually suffice. Some manufacturers, however, advise 3 or 4 units, even in relatively minor cases. In moderately severe cases, 3 to 5 units may be required; while in severe cases, 10 or more units may sometimes be needed. While as many as 45 units (450 ml. of antivenin) have been given to a single patient, this is never warranted, and indeed is very dangerous.

The choice of the route of administration will depend, among the other factors previously noted, upon the amount of time that has transpired between the bite and the administration of the antivenin. The longer the delay the more urgent the need for intravenous antivenin. However, not all products can be given intravenously with the same degree of safety. The physician should consult the brochure which accompanies the antivenin before injecting the serum. Intravenous antivenin is also indicated for those patients in shock.

In most cases, a portion of the first unit should be injected subcutaneously proximal to the bite or surrounding the wound or in advance of the swelling. *Under no circumstances should antivenin be injected into a finger or toe.* Avoid giving large amounts of the antivenin into the injured part, for this makes it difficult to determine how much swelling is due to the venom and how much is due to the presence of antivenin. A second portion of the antivenin should be injected intramuscularly into a large muscle mass distant from the wound. The last portion of the first unit should be given intravenously, if at all possible. It can be added to a physiologic saline solution and given in a continuous drip. Subsequent doses can then be added to the saline solution.

Antivenin is of value in neutralizing certain effects of the venom, but perhaps not all. It is difficult to determine how long after envenomation antivenin can be given and still be effective. Certainly, it is of value if given within 4 hours of a bite; it is of lesser value if administration is delayed for 8 hours, and it is of questionable value after 10 hours, except perhaps in cases of poisoning by certain elapids. It seems advisable to recommend its use up to 12 hours following envenomation, unless there is a definite contra-indication. Following bites by the Australian elapids it might be of value even when given beyond 12 hours following the bite.

Administration of antivenin is not a procedure without danger. In sensitive persons its injection can be fatal. In persons with a history of extensive allergies it must be injected with extreme caution, even in the presence of a negative skin test. Approximately 80 percent of one large group of Americans tested for horse serum sensitivity following rattlesnake bites had negative or only slightly positive reactions. Twenty percent of this group were subsequently treated for delayed serum reactions; reactions were most marked in those patients receiving 3 vials or more of antivenin.

In patients sensitive to horse serum, desensitization should be carried out as indicated in the brochure accompanying the antivenin, or according to standard medical procedures for desensitization. In those patients having a history of sensitivity and a strongly positive skin or conjunctival test (3 or 4+), antivenin should be withheld. However, the physician will need to weigh the risk of withholding the antivenin, against the risk of death, when poisoning has occurred by large mambas, kraits, cobras, or certain of the Australian elapids. (See p. 23 regarding use of corticosteroids.) Antivenin has been given to very sensitive patients in a slow drip of physiologic saline, but only in a hospital where systemic arterial and venous pressures and respirations could be continuously monitored, and where an electrocardiogram could be watched.

Blood Transfusions and Parenteral Fluids

All severe cases of snake venom poisoning give rise, early or late in the course of the disease, to a decrease in blood flow. The shock seen immediately following the severe bite by a rattlesnake is due to the pooling of blood in the pulmonary circulation, and to a lesser extent in the larger vessels of the thorax. In such cases, the availability of blood to the heart and brain is markedly reduced, and unless circulating blood volume is restored, the patient may develop irreversible tissue changes. When shock develops late (12 to 72 hours) in the course of the disease, it is usually due to blood loss through hemorrhage. The hemorrhage may be evident in the injured

part, or it may be masked intraperitoneally or retroperitoneally, or it may occur into the gastrointestinal, urinary, or respiratory tracts. Pooling of blood in some organs may also take place and add to the decrease in circulating volume. Concomitant with these changes, the red blood cells may undergo lysis and further embarrass the circulation. To combat these effects, blood volume and blood flow must be maintained.

Parenteral fluid should always be given following a severe envenomation. It may be necessary to add a vasopressor drug to the solution. Avoid using corticosteroids, particularly if antivenin has or is being administered. While plasma or plasma expanders can be given, whole blood should be administered if it is available. In cases of crotalid and viperid venom poisoning, fresh blood is preferred, as the patient may be unable to produce or circulate platelets. If, and when, bleeding begins, the hematocrit may fall rapidly necessitating a number of transfusions. Exchange transfusions should be considered when the clotting time is at infinity and the blood picture displays no evidence of improvement. As many as 25 pints of blood may need to be given to the victim of a severe rattlesnake bite.

Antibiotics

A broad-spectrum antibiotic should be given if the reaction to envenomation is severe. Since the nature of the injury predisposes to infection, and since pathogenic bacteria are likely to be introduced into the wound, the use of an antibiotic seems justified. Should infection develop, cultures and organism sensitivity tests will guide subsequent antibiotic therapy. If there is extensive skin damage, large doses of an antibiotic may be needed. In such cases, repeated wound cultures and blood counts are advisable.

Tetanus Prophylaxis

Since the members of the armed services have been routinely immunized against tetanus, a "booster shot" of tetanus toxoid should be given upon admission. The use of gas gangrene antitoxin is not warranted.

Electrolyte Balance

Because of the acute changes associated with the tissue damage produced by the venom, and the loss of blood and intracellular fluid which may occur, changes in electrolyte and fluid balance should be treated immediately.

Analgesic

Aspirin or codeine may be used to alleviate pain. Morphine may be used if the pain is severe, but should be avoided in near shock conditions or when there is a respiratory deficit. Local "blocks" with procaine and topically applied lotions or ointments are rarely effective. The affected part should be kept out of a completely dependent position so as not to accentuate pain.

Respiratory Failure

At the first signs of respiratory distress, oxygen should be given, and preparations made to apply intermittent positive pressure artificial respiration. A tracheostomy may be indicated, particularly if trismus, laryngeal spasm, and excessive salivation are present. While drugs have been given to stimulate the respiratory centers, they have not proved of particular value.

Renal Shutdown

The routine emergency measures for the treatment of renal shutdown should be followed. Shock, fluid restriction, electrolytic balance, diet, and administration of digitalis must be considered. Renal dialysis may be necessary. Peritoneal dialysis is of little value.

Sedation

Mild sedation with phenobarbital is definitely indicated in all severe bites, and where respiratory failure is not a problem. Sedation will usually reduce the amount of narcotic necessary to control the pain.

Care of the Wound

The wound should be cleansed and covered with a sterile dressing. The dressing should be changed frequently when large amounts of exu-date are present. Every attempt should be made to keep the wound and dressing dry. Avoid fasciotomy. Only when circulation is seriously threatened should a fasciotomy be done.

Other Measures

1. *Antihistamines* are of no value during the acute stages of the poisoning. They can be used subsequently to control the lesser allergic manifestations provoked by the venom or horse serum.
2. *Atropine* can be used as a parasympatholytic drug.
3. *Ammonia*, injected into the wound, is contraindicated. The injection of potassium permanganate, formaldehyde, gold salts, *et cetera*, into the injured area, or elsewhere, is of no value and should not be attempted.
4. *Corticosteroids* are probably of little value during the acute stages of viper venom poisoning, and indeed their use may be contraindicated when antivenin is being administered. They might be used as a single dose treatment for shock, if no other specific antishock drug is available. In elapid venom poisoning they appear to have found some widespread use, although the clinical evidence in support of their administration in this type of poisoning is not at present convincing. They might be used in elapid venom poisoning, but here again caution should be exercised if antivenin is to be given simultaneously.

The corticosteroids are the drugs of choice in combating any late or severe manifestations of the allergic response provoked by the venom or horse serum. In most cases these manifestations do not appear until 3 to 5 days following administration of antivenin.

5. *Cryotherapy* should be avoided. Keeping the injured part cool (40–50° F.) for several days (and the patient warm) may be of some value, but freezing the extremity or keeping it immersed in ice water for days is not recommended.
6. *EDTA* (ethylenediaminetetraacetic acid) has been suggested as an agent for combating the tissue effects produced by certain viper venom enzymes. Preliminary experimental studies have indicated that 0.025 to 0.05 molar EDTA in saline, when injected in the area of the injury, retards the development of necrosis and certain other tissue changes. There is no known contra-

indication for its use in conjunction with antivenin.

7. *Hyperbaric oxygen* has been suggested as a therapeutic measure, but has not been evaluated sufficiently to recommend its use at this time.

8. *Isolation-perfusion* of an extremity with antivenin has not been evaluated sufficiently to recommend its use at this time.

Follow-up Care

Often neglected but of the utmost importance, is the follow-up care. Contractures and amputations can be reduced by initiating corrective measures and exercises following the acute stages of the poisoning. The vesicles and necrosis should be treated in a manner similar to that advised for victims of severe burns. Within one week following the injury, or thereabouts, physical therapy should be instituted. Orthopedic consultation should be sought and a rehabilitation program arranged for the patient.

REFERENCES

BUCKLEY, E. and PORGES, N. (editors) 1956. Venoms. Amer. Assoc. Advancement Sci., Washington, D.C. 467 p.

CAMPBELL, C. H. 1964. Venomous Snake Bite in Papua and Its Treatment with Tracheotomy, Artificial Respiration and Antivenin. Trans. R. Soc. Trop. Med. Hyg. 59:263–273.

CHRISTENSEN, P. A. 1955. South African Snake Venoms and Antivenoms. South African Institute Medical Research, Johannesburg. 142 p.

EFRATI, P. and REIF, L. 1953. Clinical and Pathological Observations on Sixty-five Cases of Viper Bite in Israel. Amer. J. Trop. Med. Hyg. 2:1085–1108.

GENNARO, J. F., Jr. 1953. Observations on the Treatment of Snakebite in North America, p. 427–446. In, H. L. Keegan and W. V. Macfarlane, Venomous and Poisonous Animals and Noxious Plants of the Pacific Region. Pergamon, Oxford.

KAISER, E. and MICHL, M. 1958. Die Biochemie der Tierischen Gifts. F. Deuticke, Wien. 258 p.

KEEGAN, H. L. and MACFARLANE, W. V. (editors) 1963. Venomous and Poisonous Animals and Noxious Plants of the Pacific Region. Pergamon, Oxford. 456 p.

NATIONAL ACADEMY OF SCIENCES, NATIONAL RESEARCH COUNCIL, COMMITTEE ON SNAKEBITE THERAPY. 1963. Interim Statement on Snakebite Therapy. Toxicon 1:81–87.

PHISALIX, M. 1922. Animaux Venimeux et Venins. Masson, Paris, 2 vol.

REID, H. A. 1964. Cobra-Bites. Brit. Med. J. 2:540–545.

REID, H. A., THEAN, P. C., CHAN, K. E. and BAHARAM, A. R. 1963. Clinical Effects of Bites by Malayan Viper (*Ancistrodon rhodostoma*). Lancet 1:617–621.

RUSSELL, F. E. 1961. Use of *Crotalus* Monovalent Antivenin from Rabbit Serum. Curr. Therap. Res. 3:438–440.

RUSSELL, F. E. 1962. Snake Venom Poisoning, vol II, p. 197–210. In, G. M. Piersol, Cyclopedia of Medicine, Surgery and the Specialties. F. A. Davis, Philadelphia.

Chapter VI

RECOGNITION OF POISONOUS SNAKES

TABLE OF CONTENTS

INTRODUCTION

This chapter is designed primarily for identification of freshly killed snakes, not live snakes seen in the field, nor long preserved and faded museum specimens. Identification of live snakes in the field requires practice and experience, and the guidelines do not lend themselves to brief verbal descriptions, as a rule. It is to be hoped that the snakes submitted for identification will have their heads on and not be too badly smashed. Identification is considerably more complicated if the head is badly mutilated, and a decapitated body may be unidentifiable.

GENERAL PROCEDURES IN IDENTIFICATION

It is assumed that the user of this manual will have some knowledge of where the specimen he is trying to identify came from. For example, if a suspected coral snake is brought in for identification, there will be no reason to differentiate it from the 40 or so species of coral snakes found from Mexico southward if it is known that it was collected in North Carolina. Knowledge of the area of habitat narrows the field considerably. Identifying snakes from tropical areas often poses a problem in that tropical snake faunas are much richer in the numbers of species, and the distribution of some of these is poorly known. Nevertheless, if this manual is used correctly, and if there is an adequate specimen to work with, it should be possible to distinguish first between poisonous and nonpoisonous snakes, then, if poisonous, to ascertain the correct generic identification in about 90 percent of the cases, and finally to arrive at the correct species in about 3 out of 4 cases.

First, if there is any doubt that the animal *is* a snake and a poisonous one, or if the family of the snake is unknown, then Key to the Families of Snakes, page 30 this chapter, should be consulted. If the snake is known to be a poisonous land snake, then refer to the correct geographic section of Chapter VII and thence to the descriptions of the common species of the area; if a poisonous marine snake, refer to Chapter VIII.

If practicable every medical unit that enters an area where snakebite is a hazard should build up an identified collection of local poisonous and nonpoisonous snakes (see p. 32 for directions). Small individuals or just the heads of large snakes should be sufficient. Such collections are often essential for rapid identification of dangerous species.

If the specimen cannot be identified readily, it may be:

1. An aberrant individual or one from an atypical population;

2. An uncommon species listed in the regional table but not described in detail;

3. An unknown species or one not previously known from that geographic region;

4. A harmless species incorrectly identified as poisonous. (To confirm the family, recheck characteristics using Key to the Families, this chapter.)

In examining an unidentified snake look first at the head. In all pit vipers (family Crotalidae) there is a deep hollow between the eye and nostril and slightly below a line connecting the two (see figure 4). The impression is one of an extra nostril. (A large pit viper, *Bothrops atrox*, is known in Mexico as *cuatro narices* or four nostrils.) These pits are actually sensitive heat receptors. They absolutely identify a snake as a pit viper, since they are not seen in any other type of snake. However, some pythons and boas do have pits on the upper lip. The pits may be difficult to recognize for they are often camouflaged by the head markings so that they are not visible except by close inspection; this offers another reason for bringing the intact snake in for identification.

DISTINGUISHING FEATURES IN IDENTIFICATION

Venom Apparatus

Fangs and venom glands are the only anatomic features that set poisonous snakes apart from nonpoisonous ones. Caution is demanded in examining the mouth of a freshly killed snake; the biting reflex may persist in a severed head for as long as 45 minutes. The long, moveable fangs of vipers, normally sheathed in whitish membrane and rotated parallel to the roof of the mouth, can be readily demonstrated and recognized. Fangs of elapid snakes (cobras, kraits, mambas, and related species) are smaller in size, located toward the front of the mouth, and fixed to the jaw (see fig. 5). In cobras, mambas, and some other species they are large enough to be readily recognized, but in coral snakes and some other small elapids this is not the case. Enlarged anterior teeth are seen also in some nonpoisonous snakes and can be confusing. Sea snake fangs are small and hard to distinguish. Rear fangs in colubrid snakes are rather difficult to see and extremely difficult to differentiate from nongrooved enlarged teeth found at the back of the jaw in many nonpoisonous snakes. Fortunately only a few rear-fanged snakes in Africa are sufficiently dangerous that their identification

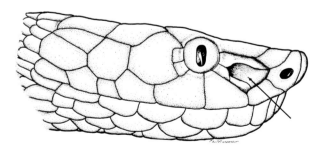

FIGURE 4.—Drawing of head of pit viper, showing the position and appearance of the loreal pit. This heat-sensitive structure is characteristic of the family Crotalidae.

is important, and the fangs in these kinds are quite long.

Head Shields

The size and arrangement of shields on the top and sides of the head are most helpful in snake identification. In the great majority of snakes

the top of the head is covered by large symmetrical shields, typically 9 in number (see fig. 6). More or less division of these shields into small scales is seen in many kinds of vipers, many boas and pythons, and in a few other kinds of snakes. Reduction of the number through fusion of shields is seen mostly in small burrowing snakes.

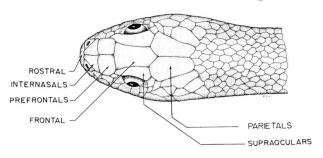

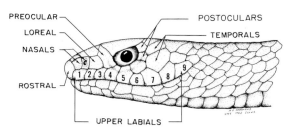

Figure 6.—Head of typical colubrid snake, illustrating arrangement of scales from dorsal and lateral views. Any of these scales may be modified in shape or absent in various groups of poisonous snakes.

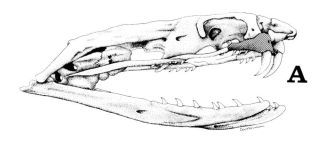

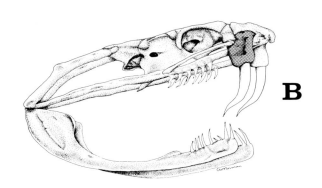

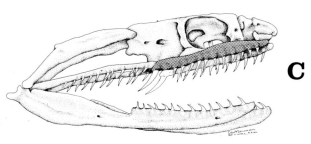

Figure 5.—Skulls representative of various families of poisonous snakes, showing lengths of maxillary bones (shaded) and positions and lengths of fangs. A. Cobra (Elapidae), showing short fang in front part of maxillary bone; B. Pit viper (Crotalidae), showing long fang on short maxillary bone; C. Rear-fanged snake (Colubridae), showing short fang on rear part of long maxillary bone (Other parts of skull diagrammatic only).

If there are typical large shields on the crown and no pit between the eye and nostril, look at the side of the head in front of the eye. The loreal shield (see fig. 6) is absent in nearly all poisonous snakes of the Elapidae as well as the African mole vipers (Viperidae). This shield is also lacking in a good many nonpoisonous snakes, but many of these are small burrowers or strictly aquatic snakes which may be eliminated on other grounds. The size of the eye may be important (see GLOSSARY).

Eye Characteristics

The shape of the pupil of the eye should be noted in live or freshly killed snakes. Most snakes have round pupils, some have vertically elliptical pupils, and a few have horizontally elliptical pupils. Vertically elliptical pupils are characteristic of most vipers but some nonpoisonous snakes also have this type. Most venomous elapids have round pupils.

Dorsal Scale Characteristics

The number of dorsal scale rows is sometimes important in snake identification. The method of counting is shown in figure 7. While it is quite possible to make this count on a snake "in the round" so to speak, the inexperienced individual may obtain better results by skinning out a section of the body and flattening the skin. It is seldom possible to take a satisfactory scale count of a live snake. It is often desirable to note if

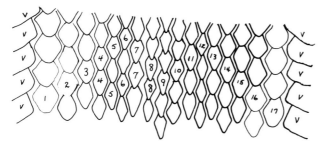

FIGURE 7.—Method of counting dorsal scale rows. Figure drawn as though skin has been slit down belly and spread flat (V = ventral plates).

the dorsal scales have a longitudinal raised ridge, *keeled*, or if they lack such ridges, *smooth* (see fig. 8).

Ventral Scutes

In the vast majority of snakes, large transverse scutes extend the full width of the belly. These are considerably reduced in size in boas and pythons, some freshwater and burrowing snakes, and in many sea snakes (see figure 9). They are completely absent in the burrowing blind snakes, and in some sea snakes. A complete count of the ventrals is routine procedure in systematic herpetology. It is easily done, but rather tedious, and is not required for most of the species identifications in this manual.

Tail

The tail of a snake begins at the anal plate which covers the opening of the cloaca. The form of the tail is often important in identification—virtually diagnostic in sea snakes and rattlesnakes. The subcaudal scutes are usually in a double row (paired); however, in some species, all or most may be in a single row (see figure 10). A count of the subcaudals is routine.

Sex

Sex of a snake can sometimes be determined readily by observing eggs or developing young in the oviducts. Pressure by fingers or injection of liquid at the base of the tail will usually evert the copulatory organs or hemipenes of a male snake. The morphology of these organs is important in snake taxonomy. Usually they are

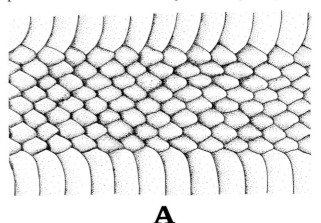

A

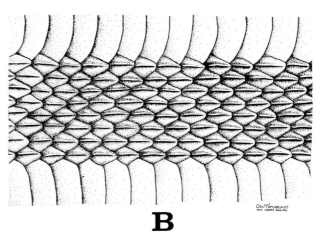

B

FIGURE 8.—Figures of dorsal scales showing major types of scale ornamentation: A. smooth scales. B. keeled scales.

A B C

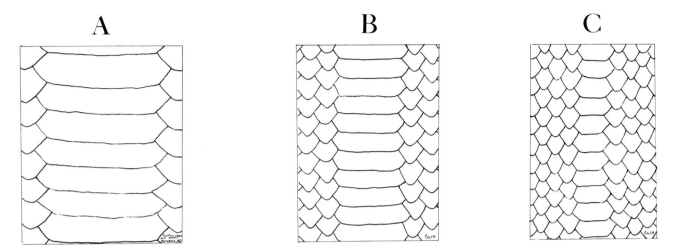

FIGURE 9.—Appearance of ventral plates in various groups of snakes. A. Extending full width of belly (most snakes); B. Moderately reduced (boas, pythons, some aquatic snakes); C. Markedly reduced (many sea snakes).

rather large fleshy structures bearing spines or other ornamentation, but they may be quite smooth, small, and slender.

Color and Pattern

Color and pattern are the most widely used but, unfortunately, are the most deceptive criteria for snake identification. Color and patterns in snakes have evolved primarily for purpose of concealment and, as a result, totally unrelated snakes may appear very much alike. Many tree snakes, for example, are green with a light line on the flank, and many snakes that live in the crevices of rock or under bark have dark heads with a light collar at the nape. Real or apparent mimicry of venomous snakes by harmless species is very widespread and may involve similarities in behavior as well as appearance. Color and pattern vary greatly even within a species. In snakes of semiarid lands, it has been observed for centuries that there is often correspondence of general body color with the color of the soil. Abnormal increases of dark pigment (melanism) or its complete absence (albinism) can in rare cases give rise to black coral snakes or white rattlesnakes. Pattern is generally more constant than color, but several kinds of snakes may show both ringed and striped types of pattern. Pat-

tern and colors of young snakes may be totally different from those of the adult. Sex differences in color and pattern are also seen.

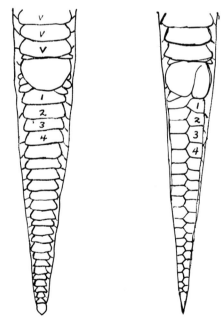

FIGURE 10.—Undersides of tails of representative snakes. Snake with ENTIRE anal plate and a SINGLE row of subcaudal scutes; snake with DIVIDED anal plate and PAIRED rows of subcaudal scutes (V = ventral plates).

29

THE FAMILIES OF SNAKES

The keys given in Chapter VII distinguish the various kinds of poisonous land snakes from one another; Chapter VIII distinguishes the poisonous sea snakes. Often, however, there are basic questions as to whether or not a snake *is* poisonous, and to what family it belongs. Sometimes a family allocation will act as a double check on a tentative identification and also, occasionally, a family designation will be all that is possible because of the condition of the snake.

The following key has been designed to sort out many kinds of nonpoisonous snakes and snake-like animals before finally distinguishing between typical harmless snakes and poisonous ones by the only positive means of identification of a poisonous kind—the presence of fangs in the upper jaw.

To identify an animal by use of this key, the reader must begin with the first couplet (pair of statements), decide which one describes the animal at hand, and then proceed to the couplet indicated at the end of the proper descriptive phrase. This procedure is followed with the next couplet and so on. Thus, an alternative decision is offered with each couplet until the reader finally determines the proper category for the animal. The animal must possess *all* of the characteristics mentioned in the proper line of couplets—not just the final characteristic. Therefore, it is always necessary to start at the beginning of the key.

The following key should always be used if there is any question as to whether or not the animal at hand is a poisonous snake:

KEY TO THE FAMILIES OF SNAKES

1. A. Body elongate, but legs or fins present on front and/or rear parts of body_____ NOT A SNAKE
   B. Body elongate, no legs or fins_____ 2

2. A. Skin slimy, with or without bony (fish-like) scales_____ NOT A SNAKE
   B. Skin dry, with thin horny scales_____ 3

3. A. Skin formed into distinct broad rings that extend around body_____ NOT A SNAKE
   B. Skin formed into small overlapping or juxtaposed scales (not in rings), at least on back_____ 4

4. A. Eye with a movable lid_____ NOT A SNAKE
   B. No movable lid_____ 5

5. A. Tail round in cross-section; not oar-shaped_____ 7
   B. Tail compressed into an oar-like blade_____ 6

6. A. Head covered with small granular scales; no large shields on crown; watersnakes of Southeast Asia_____ COLUBRIDAE
   B. Some crown shields present; see fig. 6; seasnakes, Chapter VIII_____HYDROPHIDAE ☠

7. A. A row of enlarged, transverse scutes (ventrals) down the belly_____ 11
   B. Body scales uniform above and below_____ 8

8. A. Tail with an enlarged and ornamented scute or with several spiny scales near tip (SE Asia only); Indian rough-scaled snakes_____ UROPELTIDAE
   B. No such specialized tail, a single spine or none on tip_____ 9

9. A. Eye under a distinct round scale; most of head covered with small granular scales_____ 16
   B. Eye under irregularly-shaped head plate; head covered with enlarged scutes_____ 10

10. A. Scute containing nostril forms border of lip, 14 rows of scales around body; slender blindsnakes_____LEPTOTYPHLOPIDAE
    B. Scute containing nostril separated from lip by surrounding scales; more than 14 scale rows; typical blindsnakes_____ TYPHLOPIDAE

11. A. Ventral scutes extend full width of belly_____ 15
    B. Ventral scutes narrow, not extending width of belly_____ 12

12. A. Ventrals scarcely twice size of dorsals, or less_____ 13
    B. Ventrals distinctly enlarged, more than 3 times width of dorsals; boas and pythons_____ BOIDAE

13. A. Head mainly covered with small scales_____ 16
    B. Head covered with large scutes, though not in "typical" pattern (see fig. 6)_____ 14

14. A. A large median shield behind frontal; 15 scale rows (SE Asia only); sunbeam snake_____ XENOPELTIDAE
    B. No large median scute behind frontal; 17 scale rows or more (SE Asia and northern South America). Pipe snakes_____ ANILIDAE

15. A. A spur-like hook on either side of vent (often hidden in small depressions)_____ 12
    B. No indication of spurs_____ 16

16. A. One or two large fangs near front of upper jaw on each side_____ 17
    B. No sign of fangs at front of upper jaw. Typical harmless snakes; about 2,000 species, only 2 in C. and S. Africa are dangerous_____COLUBRIDAE

17. A. Long fangs on short maxillary bone which can rotate to erect them; no other teeth on maxillary_____ 18

B. Short fangs on long maxillary bone which cannot rotate; usually teeth on maxillary bone behind fang; cobras and relatives_____ ELAPIDAE ☠

18. A. A loreal pit, see fig. 4; (SE Europe, Asia, and Americans only); pit vipers_____ CROTALIDAE ☠

B. No loreal pit (Europe, Asia, and Africa only); Old World vipers_____VIPERIDAE ☠

☠ = Families of dangerously poisonous species.

PRESERVATION AND DISPOSITION OF UNIDENTIFIED SNAKES

Snakes that cannot be identified should be preserved in the manner given in the next paragraph and submitted to the nearest U.S. Naval Preventive Medicine Unit. Such units will provide identification service. If delivery to such a unit is not practicable, then contact the nearest natural history museum or other institution which might have a staff herpetologist and request help in identification.

The two best preservatives to prepare a specimen for shipment or delivery to a herpetologist are:

1. Commercial formaldehyde diluted with 5 to 9 parts of water;

2. Grain alcohol diluted to 75 percent.

However, animals as large as most snakes will decay if placed in a preservative without some prior preparation. An ideal specimen and one which will remain in a state of minimum decay may be prepared by thoroughly injecting the body cavity and base of the tail with the preservative. A large syringe is the best means to inject the fluid, but if one is not available, multiple slits should be cut into the belly and the base of the tail and this will enable the preservative to reach the deep tissues. Then put a wad of cotton or gauze into its mouth to hold it open. The specimen should then be neatly coiled, belly side up, in a container sufficiently large to cover the snake with the preservative. Do not crowd several specimens in a single container.

Large snakes of 5 feet or more in length should be eviscerated or skinned out leaving only the head and tail intact before placing them in a container of preservative. An intact head will be sufficient to differentiate between poisonous and nonpoisonous species.

After the specimen has hardened (5 to 7 days is usually required), it may be removed from the liquid, wrapped in damp rags, put in a plastic bag and shipped to the herpetologist for identification. A tag should always be included which gives the location where the specimen was collected in enough detail so that it can be located on a map in an ordinary atlas. If the name of a small native village is used then the name of the district, department, county or other political subdivision must be added. Other information to put on the tag which will greatly increase the scientific value of a specimen includes date of collection of specimen, approximate altitude, habitat, and the name and address of the collector. Use waterproof ink or a pencil in filling out the tag.

Chapter VII

DISTRIBUTION AND IDENTIFICATION OF POISONOUS LAND SNAKES OF THE WORLD

INTRODUCTION

To facilitate the identification of some 360 species of poisonous land snakes of the world, the land areas have been divided into 10 regions (see Map 1). This chapter has been divided into 10 sections to correspond. In each section has been included a definition of the region, a list of poisonous species which occur in it and their distribution within the area and, importantly, a Key to the Genera of poisonous snakes inhabiting the region. The main body of the text of each section is separated into generic divisions (based upon the Key to Genera) and each division is headed by a description of the genus. Following thereafter are individual descriptions of the poisonous species which are responsible for the largest numbers of bites within the area, or are believed to be of serious danger to any adult human inhabiting or entering the region.

Except in a few cases which are specifically in-

dicated, the list of poisonous snakes of the world by Klemmer (1963) has been used as the basis for the nomenclature used in this chapter and elsewhere in the manual. The list of references appended to the end of each section is not intended to be comprehensive, but indicates the main sources of information utilized in preparing the accounts and may serve as an introduction to the literature on the poisonous snakes of the region. This same kind of information is given on sea snakes (Hydrophidae) in a separate chapter (Chapter VIII).

Figures have been included for all of the dangerously poisonous species if photographs or drawings were available at the time of publication. Missing photographs, or likely sources of such photographs, should be forwarded to the Preventive Medicine Division (Code 72), Bureau of Medicine and Surgery, Navy Department, Washington, D.C. for inclusion in future editions of this manual.

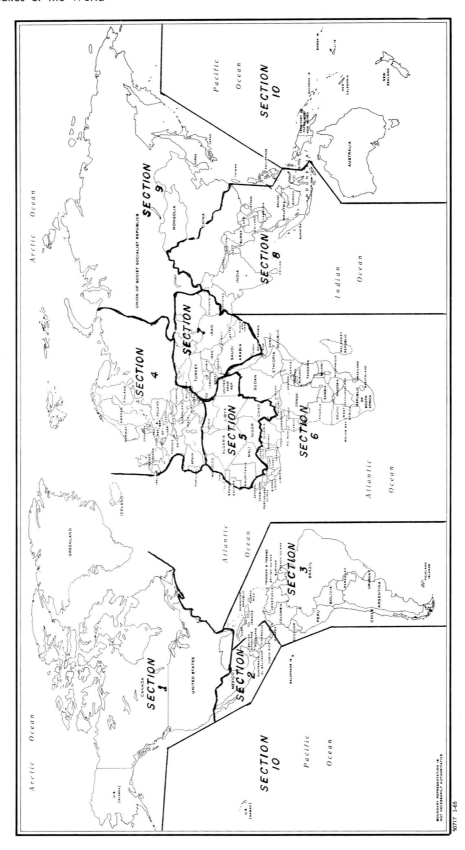

Map 1.—Map of world showing geographic divisions used in Chapter VII.

50717 1-65

BOUNDARY REPRESENTATION IS
NOT NECESSARILY AUTHORITATIVE

Section 1

NORTH AMERICA

Definition of the Region:
*Continental United States
and Canada*

TABLE OF CONTENTS

TABLE 5.—DISTRIBUTION OF POISONOUS SNAKES OF NORTH AMERICA

	New England, New York	Penn., N.J., Md., Del., Va., W. Va.	N.C., S.C., Ga., Ala., Miss., La.	Fla.	Tenn., Ky., Ill., Ind., Ohio	Ont., Mich., Wis., Minn.	Texas	Kan., Okla., Ark., Mo.	Alberta, Sask., Mont., S. Dak., N. Dak., Neb., Iowa, Colo., Utah, Wyo.	Ariz., N. Mex.	Calif., Nev.	B. Columbia, Wash., Ore., Idaho
ELAPIDAE												
Micruroides euryxanthus										S		
Micrurus fulvius			X	X			E&S	SW Ark.				
CROTALIDAE												
Agkistrodon contortrix	S	X	X	N	X		X	X	S Iowa; SE Neb.			
A. piscivorus		SE Va.	X	X	S.(Not in Ind. or Ohio)		E	E				
Crotalus adamanteus			S	X								
C. atrox							X	S Okl; SW Ark.		X	SE	
C. cerastes										SW	S	
C. horridus	X	X	X	N	X	S Wis; SE Minn.	E	E	S Iowa; SE Neb.			
C. lepidus							SW			S		
C. mitchellii										W	S	
C. molossus							W			X		
C. pricei										S		
C. ruber											SW	
C. scutulatus							SW			W	SE	
C. tigris										SW		
C. viridis						SW Minn.	W	W	X	X	X	X
C. willardi										S		
Sistrurus catenatus	W N.Y.	W Penn.			N (Not in Ky. or Tenn.)	X	X	X	Neb; Iowa; E Col.	SE		
S. miliarius			X	X	SW Tenn. & SW Ky.		E	S				

Certain groups of adjoining states are here treated as units. The symbol X indicates distribution of the species is widespread within the unit. Restriction of a species to a part of a unit is indicated appropriately (SW = Southwest, etc.).

INTRODUCTION

North America has a comparatively small but well-known poisonous snake fauna. It includes 2 species of coral snakes, the closely related copperhead and cottonmouth, and 15 species of rattlesnakes. Most of these have been further divided into subspecies so that some 39 named forms are recognized.

Poisonous snakes have been reported from all mainland states except Alaska, although they have been exterminated in Maine. Only the copperhead and 3 species of rattlesnakes have really extensive ranges. Poisonous snakes in Canada are restricted to comparatively small sections of southern British Columbia, Alberta, Saskatchewan, and Ontario.

Rattlesnakes are known from elevations up to 11,000 feet in the southern Sierra Nevada of California, to about 8,000 feet on dry, rocky slopes in Montana, and to the tops of the highest mountains in the southern Appalachians. In spite of this, poisonous snakes are rare in high mountains, in northern evergreen forests, and in heavily farmed or urban industrial areas.

Some species survive unexpectedly well in suburban areas, especially in the southern United States. Areas with unusually large populations of poisonous snakes include parts of the Great Plains (rattlesnakes), the lower Mississippi Valley and Gulf Coast (rattlesnakes and cottonmouths), and the southern Appalachians (rattlesnakes and copperheads).

Snakebite is by no means rare in the southern and western United States. Incidence is highest in children in the 5 to 15 year age group, and most bites are sustained close to home whether in rural or suburban areas. Many bites result from deliberate handling of venomous snakes. Since 1950, there have been no more than 10 to 25 deaths annually in the United States.

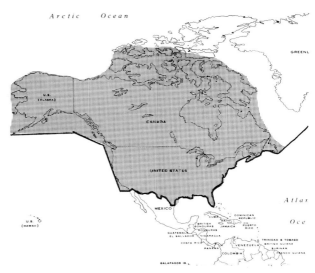

MAP 2.—Section 1, North America.

KEY TO GENERA

1. A. Loreal pit absent (see fig. 6) _____ 2
   B. Loreal pit present (see fig. 4) _____ 5
2. A. Red, black and yellow or white rings encircle the body _____ 3
   B. Ring markings not as above _____NP*
3. A. Red and yellow or white body rings in contact; end of
      snout black _____ 4
   B. Red and black rings in contact; end of snout red, white,
      yellow or black _____ NP
4. A. Yellow headband followed by black ring _____ *Micrurus*
   B. Yellow headband followed by red ring _____ *Micruroides*
5. A. Tail ends in rattle _____ 6
   B. No rattle on tail _____ *Agkistrodon*
6. A. Nine large shields on crown _____ *Sistrurus*
   B. Crown shields small or fragmented into scales _____ *Crotalus*

* NP = Nonpoisonous

GENERIC AND SPECIES DESCRIPTIONS

ELAPIDAE: Genus *Microuroides* Schmidt, 1928.

Arizona coral snake.

A single species, *M. euryxanthus* (Kennicott), is recognized. It is found in the southwestern United States and northwestern Mexico. It is a small snake but is considered dangerous (see p. 52).

Definition: Head small, not distinct from neck; snout rounded, no distinct canthus. Body slender and elongate, not tapered; tail short.

Eyes small; pupils round.

Head scales: The usual 9 on the crown. Laterally, nasal in contact with single preocular. Ventrally, mental separated from anterior chin shields by first infralabials.

Body scales: Dorsal smooth, in 15 nonoblique rows throughout body. Ventrals 206–242; anal plate divided; subcaudals paired, 19–32.

Maxillary teeth: Two relatively large tubular fangs followed, after an interspace, by 1–2 small teeth.

Remarks: Differs from nonpoisonous snakes as *Micrurus* does; differs from *Micrurus* in the solid black head color which ends in a straight line across the parietals, and in the teeth behind the fangs.

ELAPIDAE: Genus *Micrurus* Wagler, 1824.

American coral snakes.

About 40 species are currently recognized. They range from North Carolina to Texas, and from Coahuila and Sonora, Mexico, southward through Central and South America to Bolivia and Argentina. Most are small species but some attain lengths in excess of 4 feet. All are dangerous.

Definition: Head small, not distinct from neck; snout rounded, no distinct canthus. Body elongate, slender, not tapered; tail short.

Eyes small; pupils round.

Head scales: The usual 9 on the crown, Laterally, nasal in contact with single preocular. Ventrally, mental separated from anterior chin shields by first infralabials.

Body scales: Dorsals smooth, in 15 nonoblique rows throughout body. Ventrals 177–412; anal plate divided or entire; subcaudals 16–62, usually paired but more than 50 percent single in some species.

Maxillary teeth: Two relatively large tubular fangs with indistinct grooves; no other teeth on bone.

Remarks: Nearly all coral snakes have color patterns made up of complete rings of yellow (or white), black, and usually red.

Eastern Coral Snake, *Micrurus fulvius* (Linnaeus).

Identification: Head small; body slender with little taper; tail short; scales smooth with high gloss.

End of snout black followed by broad yellow band across base of head and wide, black, neck ring. Body completely encircled by black, yellow, and red rings—*the red and yellow rings touching*. If the red and black rings touch each other, if the end of the snout is red, whitish, or speckled, and if the colors fail to encircle the body, the snake is not a North American coral snake (see plate II, fig. 5). *These rules are not necessarily true in tropical America.* In the small Arizona coral snake (*Microuroides euryxanthus*) the yellow head band is followed by a wide red neck ring (see fig. 25).

Average length 23 to 32 inches; maximum 47 inches.

Distribution: Southern United States from coastal North Carolina to west Texas and into northeastern Mexico at low elevations. Inhabits grassland and dry open woods; sometimes found along streams; occasionally in suburban areas.

Remarks: Very secretive but sometimes found in the open during early or midmorning. Rather quick in its movements. When restrained it elevates the tail with the tip slightly curled and frequently tries to bite.

Venom of this coral snake is very toxic but small in quantity. Many bites seem to be ineffective. In a recently reported series of 20 cases, 10 showed little or no evidence of poisoning. However, of 6 that showed definite signs of systemic envenomation, 4 died. A species specific antivenin soon will be available from Wyeth Laboratories.

CROTALIDAE: Genus *Agkistrodon* Beauvois, 1799.

Moccasins and Asian pit vipers.

Twelve species are recognized. Three of these are in North and Central America; the others are in Asia, with one species, *A. halys* (Pallas) ranging westward to southeastern Europe. The American copperhead (*A. contortrix*) and the Eurasian mamushi and its relatives (*A. halys*) seldom inflict a serious bite but *A. acutus* and *A. rhodostoma* of southeastern Asia, as well as the cottonmouth (*A. piscivorus*) of the southeastern United States, are dangerous species.

Definition: Head broad, flattened, very distinct from narrow neck; a sharply-distinguished canthus. Body cylindrical or depressed, tapered, moderately stout to stout; tail short to moderately long.

Eyes moderate in size; pupils vertically elliptical.

Head scales: The usual 9 on the crown in most species; internasals and prefrontals broken up into small scales in some Asian forms; a pointed nasal appendage in some. Laterally, loreal pit separated from labials or its anterior border formed by second supralabial. Loreal scale present or absent.

Body scales: Dorsals smooth (in *A. rhodostoma* only) or keeled, with apical pits, in 17–27 nonoblique rows. Ventrals 125–174; subcaudals single anteriorly or paired throughout, 21–68.

American Copperhead, *Agkistrodon contortrix* (Linnaeus).

Identification: Head triangular; body moderately stout; facial pit present; pupil elliptical; most of subcaudals undivided.

Pinkish-buff, russet, or orange brown with dark brown to reddish crossbands; belly pinkish white with large dark spots or mottling; top of head yellowish to coppery-red; sides paler; end of tail yellow in young, black to dark greenish or brown in adult. The crossbands are narrow in the center of the back and wide on the sides in eastern specimens, only slightly narrowed in western ones (see plate I, fig. 5; plate III, fig. 6; plate IV, fig. 1).

Average length 2 to 3 feet; maximum slightly over 4 feet; males larger than females.

Distribution: The eastern United States (Massachusetts to Kansas and southward exclusive of peninsular Florida), westward into trans-Pecos, Texas. Frequents wooded, hilly country in the north and west; lowlands in the south; sometimes plentiful in well populated areas.

Remarks: Nocturnal in warm weather, diurnal in cool. In rocky country frequently hibernates in ledges with rattlesnakes and various nonpoisonous species. Usually remains coiled and quiet unless closely approached or touched; vibrates tail when angry; often seems reluctant to strike, but some individuals are very irritable.

Copperheads account for the great majority of snake bites seen in the eastern United States, exclusive of Florida and the Mississippi delta. Fatalities are almost unknown.

Cottonmouth, *Agkistrodon piscivorus* (Lacépède).

Identification: A pit viper related to the copperhead but very widely confused with nonpoisonous semiaquatic snakes of the genus *Natrix*. For identification of dead specimens, note presence of facial pit, elliptical pupil, undivided subcaudals—all features lacking in nonpoisonous snakes within the range of the cottonmouth. For field identification, head of cottonmouth is decidedly heavier and eyes less prominent than in the harmless water snakes. Behavior further distinguishes it (see Remarks).

Olive or brown with wide blackish crossbands often

FIGURE 11.—Cottonmouth. *Agkistrodon piscivorus*. Photo by Isabelle Hunt Conant. (See also plate I, fig. 6; plate III, fig. 5)

enclosing lighter centers; belly is yellow and heavily marbled with black or dark gray; dark stripes behind eye; end of tail black. Large snakes, especially in the western part of the range, may be almost uniformly black above. Young have a more vivid pattern and a yellowish tail.

Average length 30 to 45 inches; maximum about 6 feet.

Distribution: Southeastern Virginia through southern lowlands and up Mississippi valley to southern Illinois; west to central Texas, the southern third of Missouri, and extreme southeastern Kansas. Inhabits swamps, shallow lakes, and sluggish streams; usually absent from swift, deep, cool water.

Remarks: Often seen basking by day on logs, stones, or branches near water; also active at night in warm weather. Frequently it is a belligerent snake that does not try to escape but throws back its head with mouth widely open showing the white interior and at the same time twitching or vibrating the tail. Nonpoisonous water snakes almost always swim or crawl away rapidly when alarmed.

Bites by cottonmouths are fairly frequent in the lower Mississippi Valley and along the Gulf Coast. Fatalities are rare, but the venom has strong proteolytic activity. Tissue destruction may be severe. There is no species specific antivenin for the cottonmouth and copperhead. Polyvalent Crotalid Antivenin (Wyeth Inc., Philadelphia) should be used.

RATTLESNAKES

Rattlesnakes are distinctively American serpents that can be almost always identified by the jointed rattle at the tip of the tail. The rattle is vestigial in a single rare species found on an island off the Mexican coast. It is too small to be a good field identification characteristic in the pigmy rattlesnakes (*Sistrurus miliarius*) and in young of some other small rattlesnakes. Although most of the rattle can easily be pulled or broken off, the base or matrix usually remains. Rarely the entire tail tip including the rattle matrix may be missing as a result of injury. Nine large crown shields are seen in rattlesnakes of the genus *Sistrurus*. In the genus *Crotalus* the crown shields are more or less extensively fragmented. The facial pit is present in all rattlesnakes (see fig. 4 p. 26). Scales are keeled and subcaudals undivided.

Species identification among rattlesnakes may be difficult, but it is often important. The venoms show significant differences that can influence treatment and prognosis. Polyvalent Crotalid Antivenin (Wyeth, Inc., Philadelphia) is specific for the venoms of the eastern and western diamondbacks (*Crotalus adamanteus* and *C. atrox*). It is effective to some degree against all rattlesnake venoms.

The larger species of rattlesnakes feed principally upon small mammals; the smaller species mostly upon lizards. All rattlesnakes are live-bearing.

CROTALIDAE: *Genus Crotalus Linnaeus, 1758.*

Rattlesnakes.

Genus *Crotalus* Linnaeus, 1758. Rattlesnakes.

About 25 species of rattlesnakes are currently recognized. Most species are in the southwestern United States and northern Mexico. One specie's (*C. durissus*) ranges southward into southern South America, two are found east of the Mississippi River, and two as far north as Canada. A few of the very small species, and small individuals of large species (less than 2 feet) may offer little danger, but most species do; some are highly dangerous.

Definition: Head broad, very distinct from narrow neck, canthus distinct to absent. Body cylindrical, depressed, or slightly compressed, moderately slender to stout; tail short with a horny segmented rattle.

Eyes small; pupils vertically elliptical.

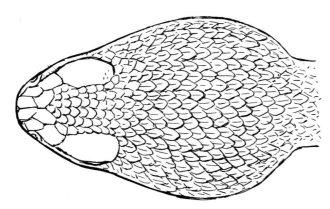

FIGURE 12.—Head of Eastern Diamondback Rattlesnake, *Crotalus adamanteus*, showing absence of many crown scutes. Drawing by Lloyd Sandford.

Head scales: Supraoculars present, a pair of internasals often distinct, occasionally a pair of prefrontals; enlarged canthal scales often present; other parts of crown covered with small scales. Laterally, eye separated from supralabials by 1–5 rows of small scales.

Body scales: Dorsals keeled, with apical pits, in 19–33 nonoblique rows at midbody. Ventrals 132–206; subcaudals 13–45, all single or with some terminal ones paired.

Eastern Diamondback Rattlesnake, *Crotalus adamanteus* Beauvois.

Identification: Within its range the only large rattlesnake with distinct, diagonal, whitish stripes on side of head; tail more or less indistinctly ringed.

Olive green to dark brown with central series of darker diamond shaped blotches each with a somewhat lighter center and a distinct cream or yellow edge; belly cream heavily clouded with gray.

Average length 3½ to 5½ feet; maximum 8 feet.

FIGURE 13.—Eastern Diamondback Rattlesnake, *Crotalus adamanteus*. Photo by Isabelle Hunt Conant.

Distribution: Coastal lowlands from North Carolina through Florida to extreme eastern Louisiana. Found in dry pine woods, palmetto thickets, old fields. However, may enter water, either fresh or salt.

Western Diamondback Rattlesnake, *Crotalus atrox* Baird and Girard.

Identification: Two light, diagonal stripes on side of head, posterior one extending to angle of mouth; tail distinctly ringed with black and gray or white, the black rings as wide as or wider than the pale ones; scales

FIGURE 14.—Western Diamondback Rattlesnake, *Crotalus atrox*. Photo courtesy *Scientific American*. (See also plate III, fig. 2).

between supraoculars small; two scales between nasals and in contact with rostral.

General coloration buff, gray, brown, or reddish with diamonds that are less clear-cut, often appearing dusty with indistinct light edges; belly cream to pinkish buff sometimes clouded with gray.

Average length 3 to 5½ feet; maximum 7 feet.

Distribution: Central Arkansas to southeastern California southward through most of Texas into Mexico to northern Veracruz and southern Sonora with an isolated population in Oaxaca. Inhabits many types of terrain from dry, sparsely wooded rocky hills to flat desert and coastal sand dunes. Often found in agricultural land and near towns. Generally avoids dense forest, swamps, and elevations above 5000 feet in the United States but may be found up to 8,000 feet in Mexico.

Red Diamond Rattlesnake, *Crotalus ruber* Cope.

Identification: Separated from the western diamondback only by its more reddish color and minor details of scalation (usually 29 rather than 25 scale rows at midbody; first lower labial usually divided in *ruber*, undivided in *atrox*).

Average length 40 to 50 inches; maximum about 5 feet.

FIGURE 15.—Red Diamond Rattlesnake, *Crotalus ruber*. Photo by Isabelle Hunt Conant.

Distribution: Baja California and southwestern California; this species and *C. atrox* meet only in a narrow zone in extreme northeastern Baja California and adjacent California. Red diamond rattlesnakes are largely confined to rocky hillsides with scrubby vegetation but at no great elevation.

Remarks: These 3 large rattlesnakes, the red diamond, eastern diamondback, and western diamondback, are quite similar in appearance but differ somewhat in behavior. The red diamond rattler is the most diurnal of the group, although all may be active by day during cooler times of the year. Western diamondbacks in the northern part of their range aggregate in large numbers to hibernate—a trait seen in some other species of rattlesnakes as well.

Temperament in the group shows much individual variation. Generally the red diamond rattler is the mildest mannered, the western diamondback the most irritable. All may defend themselves with great vigor. They sometimes raise the head and a loop of the neck well above their coils to gain elevation in striking. All may occasionally strike without rattling. The red diamond rattler often hisses loudly.

These snakes have long fangs and copious venom; the bite of an adult of any of the 3 is a serious matter. The eastern and western diamondbacks cause most of the snakebite fatalities in the United States. Venom of the red diamond rattlesnake is definitely less toxic and fatalities from its bite are rare.

Mojave Rattlesnake, *Crotalus scutulatus* (Kennicott).

Identification: Very similar to the western diamondback and prairie rattlesnakes in pattern and general appearance. Scales on top of head between and anterior to eyes large, resembling shields of most snakes; dark rings on tail much narrower than light spaces between them; general color often greenish or olive.

Average length 30 to 40 inches; maximum about 4 feet.

FIGURE 16.—Mojave Rattlesnake, *Crotalus scutulatus*. Photo by Isabelle Hunt Conant.

Distribution: West Texas northwestward to the Mojave Desert of California and southeastward on the Mexican highland. Occurs very largely in desert and prairie-desert transition zone. Decidedly a lowland snake in the northern part of its range; frequents arid mountain country in Mexico.

Remarks: Habits much like those of the western diamondback but not generally so bad tempered.

It is important to recognize the Mojave rattlesnake, for its venom is more toxic and has a more marked effect on respiration than that of any other North American rattlesnake. Bites by this species oftentimes show little local reaction and may not be considered serious until severe respiratory difficulty supervenes.

Rattlesnakes of the *Crotalus viridis* complex are widespread and sometimes difficult to differentiate. However, over a large part of their range they are the only rattlesnakes present. The following characteristics are helpful in identification:

1. Light diagonal stripes on side of head, if present, extend to behind angle of mouth;

2. Tail may be about half dark or ringed. If ringed, the light color is that of the body;

3. Pattern usually of crossbands or spots rather than diamonds;

4. Usually 4 scales between nasals and in contact with rostral.

Prairie Rattlesnake, *Crotalus v. viridis* (Rafinesque).

FIGURE 17.—Prairie Rattlesnake, *Crotalus v. viridis.* Photo by L. M. Klauber. (See also plate III, fig. 3.)

Identification: Light diagonal stripe behind eye narrow; body blotches rectangular, usually with narrow light edges; ground color often greenish-gray or olive-brown.

Average length 3 to 4 feet; maximum a little under 5 feet; males larger than females.

Distribution: South Dakota, Nebraska, and Kansas west to about the Continental Divide; north into southern Canada, south into extreme northern Mexico. Occurs in dry grassland and rocky hills; on open rocky mountain slopes to at least 9,000 feet.

Great Basin Rattlesnake, *Crotalus v. lutosus* Klauber.

FIGURE 18.—Great Basin Rattlesnake, *Crotalus viridis lutosus.* Photo by New York Zoological Society.

Identification: Light stripe behind eyes wider; pattern of crossbands usually without light edges; ground color buff or drab yellow.

Average length 24 to 35 inches; maximum 50 inches.

Distribution: Western Utah, southern Idaho, Nevada, southeastern Oregon. Frequents arid to semiarid rocky areas.

Pacific Rattlesnake, *Crotalus v. oreganus* Holbrook.

Identification: Light stripe behind eye wide, often indistinct; pattern of diamonds or hexagonal blotches; ground color dark gray, olive or brown. *Crotalus v. helleri*, a southern subspecies, differs from *oreganus* only in minor details.

Average length 3 to 4 feet; maximum a little over 5 feet.

Distribution: Southern British Columbia, western Washington, most of Oregon and the northern two-thirds of California, southern California mostly west of

FIGURE 19.—Southern Pacific Rattlesnake, *Crotalus viridis helleri.* Photo by Findlay E. Russell.

the coast range, the northern half of Baja California. (Composite range for *v. oreganus* and *v. helleri*). Absent from the humid Pacific Northwest and largely confined to semiarid regions in Washington and Oregon. Common over much of California from sea level to 11,000 feet but avoids extreme desert conditions. May be plentiful in agricultural districts and suburbs.

Remarks: Rattlesnakes of the *Crotalus viridis* group are largely diurnal, although they avoid intense light and heat. In the northern part of their range they assemble in great numbers to hibernate.

In disposition these snakes are, on the average, less irritable than diamondbacks and less likely to make a determined defense. A characteristic defensive gesture is to protrude the tongue as far as possible and wave it slowly up and down.

Bites from rattlesnakes of the *viridis* group are relatively common. There is evidence that venom of the Pacific subspecies *oreganus* and *helleri* is more toxic than that of eastern subspecies. Numbness and prickling sensations about the mouth are rather common; local symptoms may not be proportionately severe. Bites of the small Great Basin and Colorado Plateau subspecies rarely are dangerous.

Sidewinder, *Crotalus cerastes* Hallowell.

Identification: Presence of an elevated hornlike scale above the eye identifies this rattlesnake.

Cream, tan, gray, light brown or pinkish with rows of darker spots; tail ringed.

Average length 18 to 25 inches; maximum about 30 inches; females slightly larger than males.

Distribution: Deserts of southeastern California and southern Nevada southward through western Arizona into adjacent Sonora and Baja California. Most common in sandy flats and dunes with sparse vegetation; sometimes on arid rocky hillsides.

Remarks: Sidewinders often rest during the day with part of their body buried under sand and are active at night. The sidewinding type of motion, difficult to describe but unmistakable when seen, is characteristic of this snake and some heavy-bodied sand vipers of Africa and Asia. It is used occasionally by some other desert snakes including a few other species of rattlesnakes. The name sidewinder is also applied incorrectly to other kinds of small rattlesnakes in the southwestern United States.

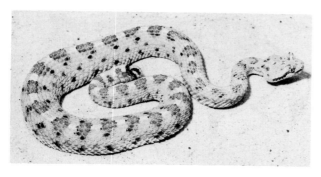

Figure 20.—Sidewinder, *Crotalus cerastes*. Photo by New York Zoological Society.

The disposition of the sidewinder is about the same as that of the *viridis* group of rattlesnakes. Bites, formerly quite unusual, have become more frequent with the growing use of desert areas for residential and recreational purposes. Fatalities from bites are few because the quantity of venom is small.

Timber Rattlesnake, *Crotalus horridus* Linnaeus.

Identification: The only rattlesnake of the eastern United States showing the combination of small scales between the eyes, no prominent light stripes on the side of the head and, in adult snakes, a black tail.

Yellow, gray, buff, or pale brown with sooty black crossbands or chevrons narrowly edged with pale yellow or white; often an amber, pinkish or rusty stripe down the middle of the back; belly cream to pinkish white more or less suffused with dark gray. Specimens from upland areas of the eastern United States are sometimes almost uniformly black above.

Average length 3 to 4 feet; maximum a little over 6 feet.

Figure 21.—Timber Rattlesnake, *Crotalus horridus*. Photo by New York Zoological Society. (See also plate III, fig. 4.)

Distribution: New England to the Florida panhandle, west to central Texas, north in the Mississippi Valley to southeastern Minnesota. Found in wooded rocky hills in the northern part of the range, swamps and lowland forest in the south.

Remarks: Timber rattlesnakes congregate in numbers to bask and hibernate in rocky bluffs and ledges— a habit which has greatly facilitated their extermination in populous areas. They are secretive and partly nocturnal during hot weather.

Rather mild tempered, they often do a good deal of preliminary rattling and feinting before striking. This rattlesnake and the copperhead are used in rituals by the snake-handling cults of the southern mountains. Bites among the cultists are fairly frequent, and no medical care is given as a rule. At least 20 fatalities have occurred among these snake handlers.

CROTALIDAE: Genus *Sistrurus* Garman, 1883.

Pigmy Rattlesnakes.

Three species are recognized; two are in the eastern and central United States, the other in the southern part of the Mexican plateau. None is considered especially dangerous, although *S. catenatus* is reported to sometimes cause death in children.

Definition: Head broad, very distinct from narrow neck; canthus obtuse to acute. Body cylindrical, tapered, slender to moderately stout; tail short, terminating in a relatively small horny, segmented rattle.

Eyes small to moderate in size; pupils vertically elliptical.

Head scales: The 9 typical scales on the crown. Laterally, nasal in contact with upper preocular or separated from it by loreal scale; eye separated from supralabials by 1–3 rows of small scales.

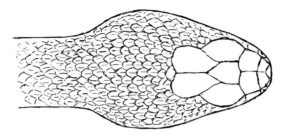

FIGURE 22.—Head of Massasauga, *Sistrurus catenatus*, showing appearance of large crown scutes characteristic of this genus. Drawing by Lloyd Sandford.

Body scales: Dorsals strongly keeled, with apical pits, in 19–27 nonoblique rows at midbody, fewer anteriorly and posteriorly. Ventrals 122–160; subcaudals 19–39, all entire or a few posterior ones paired.

Remarks: Brattstrom (1964) suggested that the genus *Sistrurus* was not recognizable, and that the 3 included species should be placed with the other rattlesnakes in the genus *Crotalus*.

Pigmy Rattlesnake, *Sistrurus miliarius* (Linnaeus).

Identification: This species and the Massasauga (*S. catenatus*) are the only United States rattlesnakes with the crown covered by large shields. In this species the tail is relatively long and slender, terminating in a tiny rattle that may be difficult to see under field conditions.

Ground color light gray, tan, reddish-orange or dark gray often with an orange or rusty midline stripe; 5 rows of sooty spots or short dark crossbars; belly white heavily clouded and spotted with black; tail barred.

Average length 15 to 22 inches; maximum 31 inches.

FIGURE 23.—Pigmy Rattlesnake, *Sistrurus miliarius*. Photo by New York Zoological Society.

Distribution: The southern lowlands and piedmont

from North Carolina to east Texas and north to southern Missouri. Frequents pine woods and grassy marshes in the southern part of the range; low rocky wooded hills in northwest.

Remarks: The rattle of these snakes is audible only at very close range. They are rather alert and bad tempered. The bite can be followed by severe pain and extensive swelling even when the snake is a small one only 6 to 9 inches long. No well documented fatal case is on record, however.

Massasauga Rattlesnake, *Sistrurus catenatus* (Rafinesque).

Identification: The large shields of the crown distinguish this species from all other United States rattlesnakes except the pigmy rattlesnake. It is best differentiated from that species by the shorter tail and well developed rattle. Ranges of the two overlap only in small areas of Texas and Oklahoma.

Ground color gray, tan, buff or yellowish with rows of dark gray, brown or black spots often with narrow light edges; belly marbled with dark gray, black and white; tail barred. Specimens from the northeastern part of the range sometimes are uniformly black when adult.

Average length 18 to 28 inches; maximum 37 inches; males larger than females.

FIGURE 24.—Massasauga, *Sistrurus catenatus*. Photo by New York Zoological Society. (See also plate I, fig. 2.)

Distribution: The Great Lakes region southwestward to extreme southeastern Arizona and southern Texas; presumably in adjacent northern Mexico. Inhabits bogs and marshes in the northeast, prairie in the west and southwest.

Remarks: Highly secretive snakes usually remaining quiet or seeking to escape when encountered but biting readily when angered. The venom is highly toxic for experimental animals, and there have been recent well authenticated cases of fatal bites in man.

REFERENCES

CONANT, Roger 1958. A Field Guide to Reptiles and Amphibians of the United States and Canada East of the 100th Meridian. Houghton Mifflin: Boston, xiii + 336 pp., pls. 1–40 (some color), figs. 1–62, Maps 1–248.

FITCH, Henry S. 1960. Autecology of the Copperhead. Univ. Kansas Publ. Mus. Nat. Hist., vol. 13, pp. 85–288, pls. 13–20. figs. 1–26, tables 1–26.

KLAUBER, LAURENCE M. 1956. Rattlesnakes: Their Habits, Life Histories, and Influence on Mankind. Univ. of California Press: Berkeley and Los Angeles. Vol. I, xxix + 1–708 pp., figs., tables, maps, frontis. (color). Vol. II, xvii + 709–1476 pp., figs., tables, frontis., (color).

STEBBINS, Robert C. 1954. Amphibians and Reptiles of Western North America. McGraw-Hill: New York, xxii + 1–528 pp., pls. 1–104.

WRIGHT, Albert H. and Anna A. WRIGHT 1957. Handbook of Snakes of the United States and Canada. Comstock: Ithaca, N.Y., 2 vols., 1105 pp., 305 figs., 70 maps.

NOTES

Section 2

MEXICO AND CENTRAL AMERICA

Definition of the Region:

All of Mexico, British Honduras, Guatemala, El Salvador, Honduras, Costa Rica, and Panama. The islands just offshore, but not the islands of the West Indies, are included.

TABLE OF CONTENTS

TABLE 6.—DISTRIBUTION OF POISONOUS SNAKES OF MEXICO & CENTRAL AMERICA

	Mexico	B. Honduras	Guatemala	El Salvador	Honduras	Nicaragua	Costa Rica	Panama
ELAPIDAE								
Micruroides euryxanthus	NW							
Micrurus alleni						X	X	X
M. ancoralis								S
M. bernadi	W							
M. browni	S							
M. clarki			X				X	X
M. diastema	E	X	E		W			
M. dissoleucus								S
M. distans	W							
M. elegans	E		X					
M. ephippifer	W							
M. fitzingeri	E							
M. fulvius	NE							
M. laticollaris	SW							
M. latifasciatus	S		S					
M. mipartitus	S					X	X	X
M. nigrocinctus	S	X	S	X	X	X	X	X
M. nuchalis	S							
M. ruatanus					X			
M. stewarti								X
CROTALIDAE								
Agkistrodon bilineatus	X	X	X	X	X	X	X	
Bothrops atrox	X	X	X	X	X	X	X	X
B. barbouri	SW							
B. bicolor	S		S					
B. dunni	S							
B. godmanni	S	?	X	X	X	X	X	X
B. lansbergii	S	?	X	X	X	X	X	E
B. lateralis							X	
B. melanurus	S							
B. nasutus	E		X	?	X	X	X	X
B. nigroviridis	S		X	?	X	?	X	X
B. nummifer	SE	X	X	X	X	X	X	X

(Continued on p. 50)

INTRODUCTION

The poisonous snakes of northern Mexico and of the Mexican plateau southward to Mexico City are very similar to those of the United States. These high and arid regions are inhabited mainly (speaking in terms of poisonous snakes) by various species of rattlesnakes. However, the Arizona coral snake (*Micruroides euryxanthus*) also is found along the northern border of the western states of Chihuahua, Sinaloa, and Sonora, and the coral snake, *Micrurus fitzingeri*, is found in the south.

As one descends from the plateau to the coastal plain, however, even as far north as Tamaulipas and Nayarit, a strange tropical snake fauna is found. Included in this are many kinds of coral snakes, the cascabel (tropical rattlesnake), *Crotalus durissus*, and various members of the American lanceheads, the genus *Bothrops*.

The coral snakes are a negligible source of danger although they are highly venomous and the case fatality rate is high (approaching 50 percent). Because they are such secretive animals, however, they are seldom encountered. Almost every coral snake bite is inflicted on a person that is attempting to catch or kill the reptile. If people would but leave the bright-colored snakes with red, black, and yellow (or whitish) rings alone, this group would offer little danger. It is the absence of a broad head and vertically elliptical pupils (characteristic of pit vipers) that causes the unknowledgeable man to mistake a coral snake for a nonvenomous species.

The other poisonous species are all pit vipers and are easily identified by the loreal pit, the broad head, eyes with vertical pupils, and the rough-scaled body. Most of the species of rattlesnakes are northern and western in distribution. Those along the Mexican–United States border are the same as those which occur in the United States.

However, the cascabel (*Crotalus durissus*) ranges through the grasslands and other dry and open areas of the tropical lowlands throughout the region as far north as southern Tamaulipas. It attains a length of 6 feet and has a large store of a very toxic venom. Apparently, too, its venom does not cause the formation of antibodies in horses to the extent that most venoms do; and,

therefore, the antivenin is only weakly effective. This makes it one of the most dangerous snakes of the region and one of the most dangerous snakes on earth.

Most of the bites through the tropical areas of Central America are inflicted by members of the American lanceheads (*Bothrops*). Many of the bites are by bush and tree vipers such as the eyelash viper (*Bothrops schlegelii*). These often cause serious injury to the affected part but seldom cause death. The major killer of man throughout the region is the barba amarilla, *Bothrops atrox* (often miscalled fer-de-lance). This 5 to 8 foot snake has an unpredictable temperament; it is easily irritated to strike and carries a large supply of powerful venom. It causes a large number of deaths each year.

The huge bushmaster (*Lachesis mutus*), on the other hand, which grows to a length of 9 to 12 feet, is seldom encountered due to its purely nocturnal habits. It causes relatively few bites, and these appear to be no more serious than those of the barba amarilla.

MAP 3.—Section 2, Mexico and Central America.

TABLE 6.—DISTRIBUTION OF POISONOUS SNAKES OF MEXICO & CENTRAL AMERICA (continued)

	Mexico	B. Honduras	Guatemala	El Salvador	Honduras	Nicaragua	Costa Rica	Panama
CROTALIDAE (continued)								
B. picadoi							X	
B. punctatus	S							X
B. schlegelii	S	X	X	X	X	X	X	X
B. sphenophrys	E							
B. undulatus	E							
B. yucatanicus	N							
Crotalus atrox	W							
C. basiliscus	W							
C. catalinensis	NW¹							
C. cerastes	W							
C. durissus	S	X	X	X	X	X	X	X
C. enyo	NW							
C. exsul	NW²							
C. intermedius	S							
C. lepidus	N							
C. mitchellii	NW							
C. molossus	X							
C. polystictus	WC							
C. pricei	NW							
C. pusillus	SW							
C. ruber	NW							
C. scutulatus	X							
C. stejnegeri	NW							
C. tigris	N							
C. tortugensis	NW³							
C. transversus	C							
C. triseriatus	X							
C. viridis	N							
C. willardi	N							
Lachesis mutus						SE	X	X
Sistrurus catenatus	N?							
Sistrurus ravus	SC							

The symbol X indicates distribution of the species is widespread within the country. Restriction of a species to part of a country is indicated appropriately (SW = southwest, etc.). The symbol ? indicates suspected occurrence of a species within the unit without valid literature.

1. Sta. Catalina Is. only   2. Cedros Is. only   3. Tortuga Is. only

KEY TO GENERA

The poisonous snakes of this region belong to two families, the Elapidae, which is represented by two genera of coral snakes, and the Crotalidae, represented by five genera. The latter are easy enough to distinguish by the presence of the loreal pit on the side of the face (fig. 4), a broad head which is distinct from a narrow neck, and the eye with a vertically elliptical pupil. However, there are several kinds of nonpoisonous snakes that look very much like the coral snakes and the latter have few easily visible features that absolutely distinguish them. Coral snakes have a relatively narrow, though often flattened, head that is not distinctly set off from the slender and cylindrical body. The eye is small and has a round pupil (as do most nonpoisonous snakes) and there is no distinctive pit on the side of the head.

In general coral snakes have rings of red, black, and yellow, but in some species the yellow may be almost white, in others the red is absent except on the head and tail and one is black and red only (brown and white in preservative). In the tricolor species the black rings may occur singly, separated from one another by rings of yellow and red, or in groups of three (triads), each triad separated by broader rings of red. The harmless mimics of these coral snakes, such as the tropical forms of the milksnake (*Lampropeltis triangulum*) and the members of such tropical genera as *Pliocercus* tend to have the black rings of their patterns *paired*. They also tend to have longer tails than the short-tailed coral snakes. However, any brightly-ringed snake should be treated with respect until its identity as a harmless species is confirmed.

1. A. Dorsal scales at midbody distinctly keeled_____ 7
   B. Dorsal scales at midbody smooth_____ 2
2. A. Pupil of eye vertically elliptical_____ NP*
   B. Pupil of eye round_____ 3
3. A. A loreal scale present (3 scales between nostril and eye)_____ NP
   B. No loreal scale_____ 4
4. A. Color pattern of body made up of alternating rings of
       red, yellow, and black (red and black in one species)_____ 5
   B. Color pattern not in rings_____ NP
5. A. Black rings alternating with yellow; OR single, sepa-
       rated by broad bands of red and yellow; OR in
       triads separated by broad bands of red_____ 6
   B. Black rings in pairs_____ NP
6. A. Entire snout and main part of head black; first band
       after yellow neck ring red_____ *Micruroides*
   B. Usually some light color anterior to eyes; first band
       after red or yellow neck band black_____ *Micrurus*
7. A. A loreal pit present_____ 8
   B. No loreal pit_____ NP
8. A. End of tail with a jointed rattle_____ 11
   B. No rattle_____ 9
9. A. Crown of head with nine regular plates_____ _____ *Agkistrodon*
   B. Crown of head with small scales or irregular plates_____ 10
10. A. Subcaudals near tip of tail divided and elongated to
        form spiny burr (Fig. 32)_____ *Lachesis*
    B. Scales near tip of tail not greatly different from those
        nearer base_____ *Bothrops*
11. A. Crown of head with nine regular plates_____ *Sistrurus*
    B. Crown of head with small scales or a few irregular plates____ *Crotalus*

_____
* NP = Nonpoisonous

GENERIC AND SPECIES DESCRIPTIONS

ELAPIDAE: Genus *Microroides* Schmidt, 1928.

Arizona coral snake.

A single species, *M. euryxanthus* (Kennicott), is recognized. It is found in the southwestern United States and northwestern Mexico. It is a small snake but is considered dangerous.

Definition: Head small, not distinct from neck; snout rounded, no distinct canthus. Body slender and elongate, not tapered; tail short.

Eyes small; pupils round.

Head scales: The usual 9 on the crown. Laterally, nasal in contact with single preocular. Ventrally, mental separated from anterior chin shields by first infralabials.

Body scales: Dorsals smooth, in 15 nonoblique rows throughout body. Ventrals 206–242; anal plate divided; subcaudals paired, 19–32.

Maxillary teeth: Two relatively large tubular fangs followed, after an interspace, by 1–2 small teeth.

Remarks: Differs from nonpoisonous snakes as *Micrurus* does; differs from *Micrurus* in the solid black head color which ends in a straight line across the parietals, and in the teeth behind the fangs.

Arizona Coral Snake, *Microroides euryxanthus* (Kennicott).

Identification: The elongate body, unmodified rostral, and black snout distinguish this species from the similarly-colored nonpoisonous sand snakes (*Chilomeniscus*) and shovel-nosed snakes (*Chionactis*) that inhabit the same region. The yellow- or white-bordered red rings distinguish it from the king snakes (*Lampropeltis*) which have black-bordered red bands. Adults average 12 to 16 inches in length; occasional individuals attain a length of 20 inches.

Snout and anterior part of head black, ending in a straight line across posterior tips of parietals. A light

FIGURE 25.—Arizona Coral Snake, *Microroides euryxanthus*. The straight line across the ends of the parietals and the red color of the first body ring are distinctive. Photo by Charles M. Bogert.

(yellow or whitish) band on neck, followed by a red ring; remainder of body with alternating rings of black and red, each separated by light rings. Tail bands alternating black and light.

Distribution: Semidesert areas from western Texas and western Chihuahua through southern New Mexico, Arizona, Sonora, and Sinaloa; on Tiburon Island. Found at altitudes up to 5,000 feet (Portal, Arizona).

Remarks: This small and secretive snake is inoffensive and very few bites have been reported. However, it possesess a highly toxic venom and should not be treated carelessly.

ELAPIDAE: Genus *Micrurus* Wagler, 1824.

American coral snakes.

About 40 species are currently recognized. They range from North Carolina to Texas, and from Mexico southward through Central and South America to

FIGURE 26.—Fitzinger's Coral Snake. *Micrurus fitzingeri* (Jan). An unusual red, yellow, and black coral snake that ranges well onto the southern part of the Mexican plateau. Photo by Charles M. Bogert.

Bolivia and Argentina. Most are small species but some attain lengths in excess of 4 feet. All are dangerous.

Definition: Head small, not distinct from neck; snout rounded, no distinct canthus. Body elongate, slender, not tapered; tail short.

Eyes small; pupils round.

Head scales: The usual 9 on the crown. Laterally, nasal in contact with single preocular. Ventrally, mental separated from anterior chin shields by first infralabials.

Body scales: Dorsals smooth, in 15 nonoblique rows throughout body. Ventrals 177–412; anal plate divided or entire; subcaudals 16–62, usually paired but more than 50 percent single in some species.

Maxillary teeth: Two relatively large tubular fangs with indistinct grooves; no other teeth on bone.

Remarks: Nearly all coral snakes have color patterns made up of complete rings of yellow (or white), black, and usually red. They differ from their non-poisonous "mimics" in that the red color, when present, is usually bordered by the yellow or white; in the non-poisonous kinds it is usually bordered in black.

Atlantic Coral Snake, *Micrurus diastema* (Duméril, Bibron, and Duméril).

Identification: A coral snake with numerous narrow black rings, which alternate with yellow and red rings. Black rings not narrowed laterally, usually complete below. Adults average 2 to 3 feet in length.

Black rings narrowly edged with yellow, which is sometimes absent; red rings of approximately the same width; red scales tipped with black. Black rings not in triads, varying from 10 in Yucatan to as many as 60 on the body in the highlands of Guatemala.

Ventrals 192–229; subcaudals 32–57; no supra-anal tubercles.

Distribution: Eastern Mexico southward through Guatemala and British Honduras to Honduras.

Remarks: This remarkably variable coral snake is fairly constant in any one region and can usually be distinguished by the irregular black spots in the red rings.

Broad-banded Coral Snake, *Micrurus distans* (Kennicott).

Identification: A coral snake with broad red bands and single narrow black bands. The head is mainly black and the lips are yellow. Adults average 2 to 3 feet in length; maximum length 42 1/4 inches.

The body color is mainly red, the red scales not black-tipped. There are 11–17 black rings on the body, the rings on the sides may be slightly narrower, and 3–6 black rings on the tail. The crown of the head is black back to the level of the eyes, but the lips are yellow (or white) and there are spots of the light color on the snout.

Ventrals 208–233; subcaudals 38–52.

Distribution: Western Mexico from Sonora to Guerrero.

Remarks: This coral snake has a remarkable harmless mimic which inhabits the same region. The neotropical milksnake, *Lampropeltis triangulum nelsoni* Blanchard, has the same broad red bands and narrow black bands. However, as in most coral snake mimics, the black bands occur in pairs—an occurrence never found in coral snakes.

Black-ringed Coral Snake, *Micrurus mipartitus* (Duméril, Bibron, and Duméril).

Identification: A coral snake with broad black rings and numerous narrow white, yellow, or red (in Central America) rings between. Adults average about 24 inches in length; occasional individuals may exceed 3 feet.

Snout black, a broad red band passing just behind eye and covering posterior part of head. Body with 34–81 black rings separated by narrow, (usually) yellow rings; tail with 3–5 black rings and 2–5 red rings.

Ventrals 197–310; subcaudals 26–34.

Distribution: Rain forest areas from Nicaragua to northern Venezuela and Peru.

Remarks: The unusual coloration of this coral snake, a red ring on the head and 2–5 others on the tail, is distinctive.

Black-banded Coral Snake, *Micrurus nigrocinctus* (Girard).

Identification: A coral snake with a black snout and broad red bands alternating with single uniform black rings, each separated from the other with relatively narrow yellow or whitish rings. Adults average 2 to 3 feet in length; occasional individuals may attain lengths of over 4 feet.

One of the coral snakes with 12–20 single black rings on the body (3–7 on tail) which are narrowly edged with yellow or whitish. Alternating red rings usually much broader than black, but relative amounts of black, yellow and red vary geographically. Snout black with darker color extending back over frontal area in a point. A broad yellow band over posterior part of head and a black ring on neck. Scales of red area often tipped with black.

Ventrals 188–240; subcaudals 31–60. Males have supra-anal tubercles.

Distribution: Lowland rain forest areas (up to an altitude of about 4,000 feet) from southern Mexico (Guerrero) southward through Central America to northwestern Colombia. This is one of the most common species of coral snakes in the region.

Remarks: Two fatal bites referrable to this species are known from Costa Rica (S. A. Minton). No antivenin is produced for this species.

CROTALIDAE: Genus *Agkistrodon* Beauvois, 1799.

Moccasins and Asian Pit vipers.

Twelve species are recognized. Three of these are in North and Central America; the others are in Asia, with one species, *A. halys* (Pallas) ranging westward to southeastern Europe. The American copperhead (*A. contortrix*) and the Eurasian mamushi and its relatives (*A. halys*) seldom inflict a serious bite but *A. acutus* and *A. rhodostoma* of southeastern Asia, as well as the cottonmouth (*A. piscivorus*) of the southeastern United States, are dangerous species.

Definition: Head broad, flattened, very distinct from narrow neck; a sharply-distinguished canthus. Body cylindrical or depressed, tapered, moderately stout to stout; tail short to moderately long.

Eyes moderate in size; pupils vertically elliptical.

Head scales: The usual 9 on the crown in most species; internasals and prefrontals broken up into small scales in some Asian forms; a pointed nasal appendage in some. Laterally, loreal pit separated from labials or its anterior border formed by second supralabial. Loreal scale present or absent.

Body scales: Dorsals smooth (in *A. rhodostoma*

only) or keeled, with apical pits, in 17–27 nonoblique rows. Ventrals 125–174; subcaudals single anteriorly or paired throughout, 21–68.

Cantil, *Agkistrodon bilineatus* Günther.

Identification: A chocolate-brown to black pit viper with typical headplates and one thin light line along the canthus, continuing behind the eye, another along the upper part of the supralabials to the corner of the mouth. Adults average 2 1/2 to 3 feet; occasional individuals exceed 4 feet.

Juvenile individuals have broad light-edged crossbands on a lighter background; these disappear in adults except for traces of the white edging which persist as narrow irregular crossbars of white-edged scales. Ventral color dark brown with black-edged white markings.

FIGURE 27.--Cantil, *Agkistrodon bilineatus*. Photo by New York Zoological Society. (See also plate I, fig. 4.)

Dorsals heavily keeled, in 23–25 rows at midbody, fewer posteriorly. Ventrals 129–144; subcaudals 59–68, the anterior 20 or so single, the posterior ones paired.

Distribution: In swampy areas and along stream banks on both coasts of Mexico and Central America from Nuevo Leon and Sonora southward to the west coast of Guatemala and the east coast of Nicaragua.

Remarks: This is the only snake within its range with the brown color and twin light stripes on the sides of the head. It is aquatic and is often found swimming. It is presumed to be a dangerous snake; it is reported to cause serious local lesions but seldom death.

CROTALIDAE: Genus Bothrops Wagler, 1823.

American lance-headed vipers.

Between 40 and 50 species are currently recognized; all are found in tropical America and southern South America. There are three general groups: 1. Large, long-tailed terrestrial species, usually with paired subcaudals; 2. Small, short-tailed terrestrial species with single subcaudals; and 3. Small to moderate-sized arboreal species with prehensile tail, most of which have at least the anterior subcaudals single. The large terrestrial species are very dangerous, the others less so.

Definition: Head broad, flattened, very distinct from narrow neck; a sharply-distinguished canthus. Body

cylindrical or moderately compressed, moderately slender to stout; tail short to moderately long.

Eyes small to moderate in size; pupils vertically elliptical.

Head scales: Supraoculars generally present, internasals often distinct, sometimes separated by small scales; remainder of crown covered with small imbricate scales; enlarged canthals sometimes present. Laterally, second supralabial may make up anterior border of loreal pit or may be separated from it. Loreal scales present or absent.

Body scales: Dorsals keeled, in 19–35 nonoblique rows at midbody. Ventrals 121–253; subcaudals single or paired, 22–83.

Barba Amarilla, *Bothrops atrox* (Linnaeus).

Identification: An olive-green, gray, or brownish snake with a pattern of lateral darker (usually) black-edged triangles whose apices meet, or nearly meet, at the vertebral line. Adults average 4 to 6 feet; record lengths exceed 8 feet.

Ground color brownish, olive, or tan, with a narrow dark postorbital stripe and a series of about 20–30 paired lateral triangles. Each marking is lighter in the center and often has a light edging to the dark-bordered triangle. Ventral surface light cream to yellow with dark blotches becoming more numerous posteriorly. Ventrals 180–220; subcaudals 46–73, all paired.

Distribution: Forest areas from southern Tamaulipas and southern Sonora, in Mexico, through all of Central America, and in South America southward to Peru and northern Brazil. A very widespread species that is common in banana, coffee, and cocoa plantations as well as in undisturbed forest regions; often found along streams.

FIGURE 28.—Barba Amarilla, *Bothrops atrox* (an individual from Trinidad). Photo by New York Zoological Society.

Remarks: This snake has long fangs and a highly toxic venom. It is probably responsible for more deaths in the Americas than any other snake. It will usually retreat if given the opportunity, but becomes aggressive if disturbed and will strike repeatedly.

Polyvalent antivenins for the bite of this snake are produced by Laboratorio Behrens (Venezuela), Instituto Butantan (Brazil), and Wyeth, Inc., Philadelphia.

Lansberg's Hognose Viper, *Bothrops lansbergii* (Schlegel).

Identification: A small brownish ground viper with upturned snout and a series of angular blotches down the back, separated into pairs by a light vertebral line. Body short and moderately stout; head broad. Adults average 18 to 24 inches in length.

Ground color light brown, tan, or gray with a dorsal series of paired dark brown blotches separated from one another by a thin light line; broadly separated from low lateral series of spots.

Canthus raised and sharp, snout raised and pointed. Eye separated from supralabials by 2–3 rows of small scales. Dorsals 25–27, heavily keeled. Ventrals 152–159; subcaudals 29–35, all single.

Distribution: In semiarid forest and brushy areas from Southern Mexico and Guatemala through Central America to Colombia and northern Venezuela.

Remarks: This is one of several hognose vipers that inhabit the dryer areas of Central and northern South America. The similar *B. nasutus* Bocourt is found from Mexico over much the same region but generally in more moist situations.

Jumping Viper, *Bothrops nummifer* (Ruppell).

Identification: A short, thick-bodied viper with dark saddle-shaped blotches on a tan or gray background. Adults average 18 to 24 inches in length.

Ground color tan, light brown or gray with about 20 dark brown or black rhomboid blotches down the back, these often connected with lateral spots to form narrow crossbands. Top of head dark with oblique postorbital band forming upper limit of light color on sides of head. Ventral color whitish, sometimes blotched with dark brown. Snout rounded, canthus sharp. Body exceedingly stout; tail short.

FIGURE 29.—Jumping Viper, *Bothrops nummifer*. With its coarse scales and diamond-shaped markings, this snake is sometimes mistaken for a young bushmaster (*Lachesis mutus*). The nonspecialized tail tip (see fig. 32) distinguishes it. Photo by New York Zoological Society.

Dorsals strongly keeled, tubercular in large individuals, in 23–27 rows at midbody fewer (19) posteri-

orly. Ventrals 121–135; subcaudals 26–36, all or mostly single. Eye separated from labials by 3–4 rows of small scales.

Distribution: Low hilly rain forest and plantations from southern Mexico to Panama.

Remarks: This is the largest of the smaller terrestrial tropical vipers. With its stout body it can strike for a distance greater than its own body length. However, it has relatively short fangs and its venom is not highly toxic.

Eyelash Viper, *Bothrops schlegelii* (Berthold).

Identification: A green, tan, or yellow tree viper with raised and pointed scales above the eye. Body moderately stout, with a prehensile tail; head broad and distinct. Adults average 16 to 24 inches in length.

Ground color green, olive-green, tan or yellow with scattered black dots which may form irregular crossbands. Green and tan individuals commonly have narrow reddish and brown crossbands or a reticulated pattern of red. Belly green or yellow, spotted with black.

FIGURE 30.—Eyelash Viper, *Bothrops schlegelii*. Photo by New York Zoological Society.

Canthus sharp; a row of small scales above eye; 2–3 of them raised and pointed. Dorsals 19–25, moderately keeled. Ventrals 138–162; subcaudals 47–62, all single.

Distribution: In trees and bushes through rain forest areas and cacao plantations from southern Mexico southward through Central America to Ecuador and Venezuela.

Remarks: There are several green "palm vipers" but *B. schegelii* is the most commonly seen and is the only one with the raised scales above the eye. None appears to be highly dangerous and no specific antivenin is produced for this group of lance-headed vipers.

CROTALIDAE: *Genus Crotalus Linnaeus, 1758.*

Rattlesnakes.

About 25 species of rattlesnakes are currently recognized. Most species are in the southwestern United States and northern Mexico. One species (*C. durissus*) ranges southward into southern South America, two are found east of the Mississippi River, and two as far north as Canada. A few of the very small species, and small individuals of large species (less than 2 feet) may offer little danger, but most species do; some are highly dangerous. Several species range into this region (see p. 50).

Definition: Head broad, very distinct from narrow neck, canthus distinct to absent. Body cylindrical, depressed, or slightly compressed, moderately slender to stout; tail short with a horny segmented rattle.

Eyes small; pupils vertically elliptical.

Head scales: Supraoculars present, a pair of internasals often distinct, occasionally a pair of prefrontals; enlarged canthal scales often present; other parts of crown covered with small scales. Laterally, eye separated from supralabials by 1–5 rows of small scales.

Body scales: Dorsals keeled, with apical pits, in 19–33 nonoblique rows at midbody. Ventrals 132–206; subcaudals 13–45, all single or with some terminal ones paired.

Mexican West-coast Rattlesnake, *Crotalus basiliscus* (Cope).

Identification: The only rattlesnake within its range with diamond-shaped dorsal markings. Body moderately stout and rather triangular in cross section. Adults average 4 to 5 feet; maximum length 6 feet, 9 3/4 inches (Klauber, 1956).

Head uniform grayish brown or olive green except for dark postorbital bar and lighter labials; no distinct markings on crown or neck. Body brown or grayish olive with 26–41 dark light-edged, rhomb-shaped (diamond) blotches. Tail gray, darker-banded or almost unicolor without distinct markings. White or cream-colored below.

Dorsals strongly keeled, in 25–29 rows at midbody, fewer posteriorly. Ventrals 174–206; subcaudals 18–36.

Distribution: The coastal plain and mountain slopes of western Mexico from southern Sonora to central Oaxaca. Mainly an inhabitant of thorn forest, but ranges upward into tropical rain forest in the south.

Remarks: Little has been reported on the effect of

FIGURE 31.—Mexican West-coast Rattlesnake, *Crotalus basiliscus*. Photo by San Diego Zoo.

the bite of this species. However, it produces large amounts of a highly toxic venom. A large individual is unquestionably a dangerous snake.

Polyvalent antivenin is produced by the Instituto Nacional de Higiene, Mexico.

CROTALIDAE: Genus *Lachesis* Daudin, 1803.

Bushmaster.

A single species, *L. mutus*, is found in tropical America. It attains a length of 9 to 12 feet and is considered dangerous.

Definition: Head broad, very distinct from narrow neck; snout broadly rounded, no canthus. Body cylindrical, tapered, moderately stout; tail short.

Eyes small; pupils vertically elliptical.

Head scales: A pair of small internasals separated from one another by small scales; a pair of narrow supraoculars; other parts of crown covered with very small scales. Laterally, second supralabial forms anterior border of loreal pit, third very large; eye separated from supralabials by 4–5 rows of small scales.

FIGURE 32.—Underside of Tail Tips of a Lancehead (*Bothrops*), above, and the Bushmaster (*Lachesis*), below. The spiny "burr" formed of divided subcaudals is distinctive of the bushmaster. Drawings by Lloyd Sandford.

Body scales: Dorsals heavily keeled with bulbous tubercles, feebly imbricate, in 31–37 nonoblique rows at midbody, fewer posteriorly. Ventrals 200–230; subcaudals mainly paired, 32–50, followed by 13–17 rows of small spines and a terminal spine.

Bushmaster, *Lachesis mutus* (Linnaeus).

Identification: This large tan or brown snake with black or dark brown rhombs is easily recognized. The peculiar burr of pointed spines near the end of the tail is distinctive. Adults average 5 to 7 feet in length; occasional individuals attain a length of 9 feet; a maximum of 12 feet has been reported.

Ground color tan or pinkish with 23–37 black or brown rhombs on body. Markings with light centers; tail dark with light crossbands. A dark postorbital stripe which continues onto the neck. White or light yellowish below.

Distribution: Rain forest and tropical deciduous forest regions from southern Nicaragua to the coastal

FIGURE 33.—Bushmaster, *Lachesis mutus.* Photo by Isabelle Hunt Conant.

lowlands of Ecuador and the Amazon basin of Peru, Bolivia, Brazil, and Paraguay.

Remarks: This is potentially a very dangerous snake with long fangs and large amounts of rather toxic venom. However, its strictly nocturnal habits keep it from coming into contact with humans except rarely and few bites have been recorded.

Specific antivenin is produced by the Instituto Butantan (Brazil).

CROTALIDAE: Genus *Sistrurus* Garman, 1883.

Pigmy Rattlesnakes.

Three species are recognized; two are in the eastern and central United States, the other in the southern part of the Mexican plateau. None is considered especially dangerous, although *S. catenatus* is reported to sometimes cause death in children.

Definition: Head broad, very distinct from narrow neck; canthus obtuse to acute. Body cylindrical, tapered, slender to moderately stout; tail short, terminating in a relatively small horny, segmented rattle.

Eyes small to moderate in size; pupils vertically elliptical.

Head scales: The 9 typical scales on the crown. Laterally, nasal in contact with upper preocular or separated from it by loreal scale; eye separated from supralabials by 1–3 rows of small scales.

Body scales: Dorsals strongly keeled, with apical pits, in 19–27 nonoblique rows at midbody, fewer anteriorly and posteriorly. Ventrals 122–160; subcaudals 19–39, all entire or a few posterior ones paired.

Remarks: Brattstrom (1964) suggested that the genus *Sistrurus* was not recognizable, and that the three included species should be placed with the other ratlesnakes in the genus *Crotalus.*

Mexican Pigmy Rattlesnake, *Sistrurus ravus* (Cope).

Identification: A small brownish rattlesnake with the 9 usual plates on the crown; the only such rattlesnake within its range. Body moderately stout; head oval. Adults average about 20 inches in length; a large individual is 24 inches.

Ground color brown or gray with 25–35 small irregular blotches down the back, small lateral spots may fuse with the dorsal row to form irregular crossbands; 6–8 dark bands on tail. Head unicolor brown or with an arrow-shaped median dark marking. Ventral surface yellowish, blotched with brown.

Dorsals moderately keeled, in 21–23 rows at midbody, fewer posteriorly. Ventrals 138–152; subcaudals 20–30.

Distribution: The southern part of the Mexican plateau.

Remarks: This is a small species and is not considered dangerous.

FIGURE 34.—Mexican Pigmy Rattlesnake, *Sistrurus ravus.* Photo by Isabelle Hunt Conant.

REFERENCES

(See also General References)

ALVAREZ del TORO, Miguel 1960. Los reptiles de Chiapas. Inst. Zool. Estado, Tuxtla Gutierrez, Chiapas, Mexico. 7–204 p., illustrated.

BURGER, W. Leslie 1950 A Preliminary Study of the Jumping Viper, *Bothrops nummifer.* Bull. Chicago Acad. Sci., 9 (3): 59–67, 1 pl.

CLARK, Herbert C. 1942 Venomous Snakes. Some Central American Records. Incidence of Snake-bite Accidents. Amer. Jour. Tropical Med., 22 (1): 37–49.

MERTENS, Robert. 1952. Die Amphibien und Reptilien von El Salvador, auf Grund der Reisen von R. Mertens und A. Zilch. Abh. Senckenbergischen Ges., 487: 1–120, 16 pls.

REFERENCES (continued)

PICADO, C. 1931. Serpientes venenosas de Costa Rica. Imprenta Alsina, San Jose, Costa Rica: 219 p., 58 figs.

SCHMIDT, Karl P. 1933. Preliminary Account of the Coral Snakes of Central America and Mexico. Zool. Ser. Field Mus. Nat. Hist., 20: 29–40.

SCHMIDT, Karl P. 1936. Notes on Central American and Mexican Coral Snakes. Zool. Ser. Field Mus. Nat. Hist., 20 (20): 205–216, figs. 24–27.

SCHMIDT, Karl P. 1941. The Amphibians and Reptiles of British Honduras. Zool. Ser. Field Mus. Nat. Hist., vol. 22, no. 8, pp. 475–510.

SCHMIDT, Karl P. 1955. Coral Snakes of the Genus *Micrurus* in Colombia. Fieldiana-Zool., 34 (34): 337–359, figs. 65–69.

SMITH, Hobart M., and E. H. TAYLOR. 1945. An Annotated Checklist and Key to the Snakes of Mexico. Bull. U. S. Natl. Mus. (187): 1–239 p.

STUART, L. C. 1963. A Checklist of the Herpetofauna of Guatemala. Misc. Pub. Mus. Zool. Univ. Michigan (122): 1–150, 1 pl., 1 map.

TAYLOR, Edward H. 1951. A Brief Review of the Snakes of Costa Rica. Univ. Kansas Sci. Bull., 34 (part 1, no. 1): 3–88, figs. 1–7, pls. 1–23.

NOTES

Section 3

SOUTH AMERICA AND THE WEST INDIES

Definition of the Region:

All of South America, including Colombia, Venezuela, British Guiana, Suri-nam, French Guiana, Brazil, Bolivia, Paraguay, Argentina, Uruguay, Chile, Peru, Ecuador, and the island of Aruba off Venezuela, in addition to those few islands of the West Indies inhabited by poisonous snakes: Martinique, Santa Lucia, Tobago, and Trinidad.

TABLE OF CONTENTS

TABLE 7.—DISTRIBUTION OF POISONOUS SNAKES OF SOUTH AMERICA & WEST INDIES

	Colombia	Venezuela	Surinam	Br. Guiana	Fr. Guiana	Brazil	Bolivia	Paraguay	Argentina	Uruguay	Chile	Peru	Ecuador	Martinique	Saint Lucia	Tobago	Trinidad
ELAPIDAE																	
Leptomicrurus collaris		X	X	X	X	NW											
L. narducci	X						X					X	E				
Micrurus albicinctus						C											
M. ancoralis	X												X				
M. annellatus			X				X					X	X				
M. averyi			X										W				
M. balzani							E										
M. bocourti	N												X				
M. carinicauda	X	X															
M. circinalis		X															X
M. corallinus						E			N								
M. decoratus						E											
M. dissoleucus	X	X															
M. dumerilii	X												E				
M. filiformis	S					N						E	E				
M. frontalis						S	X	X	N	X							
M. hemprichii	X	X	X	X	X	NE	X					X	X				
M. hollandi*	N					NE											
M. ibiboboca						NE											
M. isozonus	X	N															
M. langsdorffi	X					C	X					E	X				
M. lemniscatus	X	X	X	X	X	C	X	X	X			X	X				X
M. mertensi												NW	SW				
M. mipartitus	X	N										X	X				
M. nigrocinctus	NW																
M. ornatissimus												X	X				
M. peruvianus												X					
M. psyches	X	X	X	X	X												
M. putumayensis	X											X					
M. pyrrhocryptus									X								
M. spixii	X	X	X	X	X	X	X					X	X				
M. spurelli	X						X										
M. surinamensis	S	X	X	X		X	X					X	X				
M. tschudii		S					X					X	X				
CROTALIDAE																	
Bothrops albocarinatus													X				
B. alternatus						S		X	N	X							

*Name invalid, according to Janis A. Roze.

(Continued on p. 61)

CROTALIDAE (continued)

Species														
B. alticola							X							X
B. ammodytoides													X	X
B. andianus	X	X	X	X	N		X	X						
B. atrox	X	X	X	X	N	N	X	X	X					
B. barnetti	SE	S	?	?	C	X	X							
B. bilineatus **	X		X	X	X	X	N	X					X	
B. caribbaeus **												X		
B. castelnaudi					N			E						
B. cotiara					S		NE	E						
B. erythromelas					E			S						
B. fonsecai	S				S									
B. hyoprorus					W			X	E					
B. iglesiasi					E									
B. insularis					SE¹									
B. itapetiningae					S									
B. jararaca					S	NE								
B. jararacussu					S	X	NE			X				
B. lanceolatus **											X			
B. lansbergii	X	N		X				X						
B. lichenosus		X												
B. lojanus														
B. medusa	X	X		X				X						
B. microphthalmus	X	X		X	X	?		X	X					
B. nasutus	X	X	X	X	S			X	X					
B. neglectus			X											
B. neuwiedi					X S	E	N	SE	E					
B. peruvianus						X		X						
B. pictus		N												
B. pifanoi					E	E								
B. pirajai														
B. pulcher								E	E					
B. punctatus	X							W	W					
B. roedingeri									X					
B. schlegelii	X	X												
B. venezuelae *	?	X						X	X					
B. xanthogrammus	X	X					X							
Crotalus durissus	X	X	X	X	X	X	N	X	E					
C. vegrandis ***	X	NE	X	X	X				W					
C. unicolor	X	X	X	X	X			X	X					
Lachesis mutus	X	X	X	X	X	X	X	X	X					X

¹ (Queimada Grande Is. only)

The symbol X indicates distribution of the species is widespread within the country. Restriction of a species to part of a country is indicated appropriately. (SW = southwest, etc.). The symbol ? indicates suspected presence of a species within the unit without valid literature.

* Name invalid according to Janis A. Roze.

** See Lazell, 1964.

*** Recently given species rank by Hoge, 1965.

INTRODUCTION

The poisonous snakes of South America belong to five genera, only one of which (*Leptomicrurus*) is restricted to this continent. All but the rattle-snake genus *Crotalus*, however, are restricted to the American tropics and temperate South America.

Although the bushmaster (*Lachesis mutus*) is the largest poisonous snake of the region, it is one of the minor hazards to human life since it is mainly of nocturnal habit and is sluggish or secretive during the day. The tropical rattle-snake (*Crotalus durissus*) and the large lance-headed vipers such as *Bothrops atrox* and related species account for most of the deaths in the region.

Coral snakes (*Micrurus* and *Leptomicrurus*) are relatively common in tropical regions. About 30 species are found here. They are secretive during the day and cause relatively few cases of snakebite. However, they secrete venom of a highly poisonous, neurotoxic variety which is responsible for a very high percentage (almost 50 percent) of deaths in victims of their bites.

The islands of the Caribbean with a few exceptions are free of poisonous snakes. All of the Greater Antilles (Cuba, Jamaica, Hispaniola, and Puerto Rico) are free of poisonous kinds. Only Martinique and Santa Lucia, among the Lesser Antilles, have poisonous snakes, as do the continental islands of Margarita, Trinidad, and Tobago, and the offshore island of Aruba.

On the mainland, only the highest of the Andes are free of poisonous snakes. At least one kind of poisonous snake ranges southward onto the pampas of southern Argentina, leaving only the southernmost tip of South America and the arid plains of Chile free of venomous snakes.

KEY TO GENERA

1. A. Dorsal scales at midbody distinctly keeled_____ 7
   B. Dorsal scales at midbody smooth_____ 2
2. A. Pupil of eye distinctly vertically elliptical_____ NP*
   B. Pupil of eye round_____ 3
3. A. A loreal scale present (3 scales between nostril and eye)_____ NP
   B. No loreal scale_____ 4
4. A. Color pattern of body made up of alternating rings of
         red, yellow, and black (red restricted to head and
         tail in some)_____ 5
   B. Color pattern not in rings_____ 6
5. A. Black rings alternating with yellow; OR single, sepa-
         rated by broad bands of red and yellow; OR in
         triads separated by broad bands of red_____ *Micrurus*
   B. Black rings in pairs or single, separated by broad rings
         of red_____ NP
6. A. A yellow band across back part of head; body black
         above, with numerous crossbands of red or yellow
         below which extend up sides as triangles_____ *Leptomicrurus*
   B. Pattern not as described above_____ NP
7. A. With a loreal pit (fig. 4)_____ 8
   B. No loreal pit_____ NP
8. A. With a segmented rattle at the end of the tail_____ *Crotalus*
   B. No rattle_____ 9
9. A. Terminal subcaudals divided into short spines, forming
         a "burr" (fig. 32)_____ *Lachesis*
   B. No such burr_____ *Bothrops*

---
* NP = Nonpoisonous

GENERIC AND SPECIES DESCRIPTIONS

ELAPIDAE: Genus *Leptomicrurus* Schmidt, 1937.

Slender coral snakes.

Two species are recognized*; both are found in northern South America. These extremely elongate and slender snakes approach 3 feet in length. There are no reported bites but they are considered potentially dangerous.

Definition: Head small, not distinct from neck; snout rounded, no distinct canthus. Body extremely slender and elongate, not tapered; tail short.

Eyes small; pupils round.

Head scales: The usual 9 on the crown. Laterally, nasal in contact with single preocular. Ventrally, mental in contact with anterior chin shields.

Body scales: Dorsals smooth, in 15 nonoblique rows throughout body. Ventrals 212–410; anal plate divided; subcaudals paired, 17–35.

Maxillary teeth: Two relatively large tubular fangs; no other teeth on bone.

Remarks: These snakes differ from *Micrurus* and *Micruroides* in that the yellow crossbands are incomplete dorsally; they are best defined on the ventral surface and appear as triangles on the sides. The contact of mental and anterior chin shields also is distinctive.

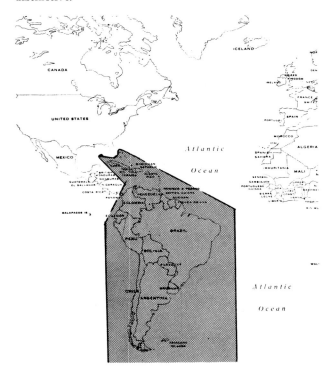

MAP 4.—Section 3, South America and the West Indies.

* A third has been described recently.

Amazon Slender Coral Snake, *Leptomicrurus narducci* (Jan).

Identification: A very elongate black coral snake with a broad yellow band on the back of the head. Adults average 24 to 30 inches; occasional individuals approach 3 feet.

Belly pattern of red (or yellow) and black crossbands, some of the red bands extending onto the sides as triangular blotches. Dorsal part of body solid black.

Ventrals 240–410; subcaudals 17–35.

Distribution: The upper Amazon region, including northwestern Brazil, eastern Ecuador, Peru, and Bolivia.

Remarks: The snakes of this genus are the only coral snakes in which the light rings are incomplete dorsally. The other species, *L. collaris* (Schlegel), differs in having fewer ventrals (212–230).

Almost nothing is known of these rare snakes. However, they attain a size that makes them a dangerous animal to pick up. No antivenin is produced for the snakes of this genus.

ELAPIDAE: Genus *Micrurus* Wagler, 1824.

American coral snakes.

About 40 species are currently recognized. They range from North Carolina to Texas, and from Coahuila and Sonora, Mexico, southward through Central and South America to Bolivia and Argentina. Most are small species but some attain lengths in excess of 4 feet. All are dangerous.

Definition: Head small, not distinct from neck; snout rounded, no distinct canthus. Body elongate, slender, not tapered; tail short.

Eyes small; pupils round.

Head scales: The usual 9 on the crown. Laterally, nasal in contact with single preocular. Ventrally, mental separated from anterior chin shields by first infralabials.

Body scales: Dorsals smooth, in 15 nonoblique rows throughout body. Ventrals 177–412; anal plate divided or entire; subcaudals 16–62, usually paired but more than 50 percent single in some species.

Maxillary teeth: Two relatively large tubular fangs with indistinct grooves; no other teeth on bone.

Remarks: Nearly all coral snakes have color patterns made up of complete rings of yellow (or white), black, and usually red.

Annellated Coral Snake, *Micrurus annellatus* (Peters).

Identification: A usually black and yellow coral snake with a narrow yellow band across the parietal scutes. This is a small species, the largest specimen is a little less than 30 inches.

Body with narrow yellow bands. Broad bands which are distinctly red in young become so darkened as to be black in most adults. This forms a pattern of alternating broad black (originally red) rings with narrower black rings. Total dark rings ranges from 37–83 on body, 5–9 on tail. The red is often visible on the belly.

Ventrals 193–225; anal plate divided; subcaudals 26–48.

Distribution: River valleys of the mountain regions of Peru, Bolivia, and Ecuador.

Remarks: This mountain species lives at altitudes of 1,500 to 6,000 feet. No reports of the effects of its bite are known.

Southern Coral Snake, *Micrurus frontalis* (Duméril, Bibron, and Duméril).

Identification: A coral snake with triads of black rings and broad red interspaces; head black with edges of plates red. Adults average 3 to 4 feet; exceptional individuals exceed 50 inches.

Crown black to the posterior end of the parietals, labials and temporals spotted with yellow, crown scutes edged with red or yellow. Body with 6–15 sets of black triads, separated with broad bands of red.

FIGURE 35.—Southern Coral Snake, *Micrurus frontalis.* The "triads" of three black and two yellow rings are characteristic of many South American coral snakes. Note that the red zones are bordered by *black* in these coral snakes. Photo by New York Zoological Society.

Ventrals; 97–230; anal plate divided; subcaudals 15–26.

Distribution: Southwestern Brazil, northern Argentina, Uruguay, Paraguay, and Bolivia.

Remarks: This is one of the larger species of coral snakes and is responsible for a number of deaths. An antivenin is prepared by the Instituto Butantan (Brazil) for this species and *M. corallinus.*

Hemprich's Coral Snake, *Micrurus hemprichii* (Jan).

Identification: A coral snake with narrow yellow and red rings, and broad black triads. Adults average 24 to 30 inches in length.

Snout and tip of chin black, with this color extending back over crown as a "cap." A red collar, narrowed

above. Body with 5–10 triads of broad black rings separated by narrow red rings.

Ventrals 159–191; anal plate entire; subcaudals 23–30.

Distribution: Rim of the Amazon basin: northeastern Brazil, the Guianas, Colombia, Ecuador, and Peru.

Remarks: This is the only species of coral snake that normally has an entire anal plate. This and the triads of broad black rings make it a distinctive snake.

Amazonian Coral Snake, *Micrurus spixii* Wagler.

Identification: A coral snake with triads of black rings which are all about equal in width and narrower than the yellow and red rings. Adults average 3 to 4 feet; occasional individuals attain a length of 5 feet.

Crown of head mainly black, often with shields edged and spotted with yellow; sides of head mostly light. Often a black collar followed by a yellow ring. Body with 4–9 complete triads of narrow and equal black rings separated by somewhat wider bands of yellow and red.

Ventrals 203–275; anal plate divided; subcaudals 16–25.

Distribution: The Amazon region; Brazil, Colombia, Venezuela, Equador, Peru, and Bolivia.

Remarks: This is one of the largest of the coral snakes, and it has been responsible for several deaths. A polyvalent coral snake antivenin is produced by the Instituto Butantan (Brazil).

Surinam Coral Snake, *Micrurus surinamensis* (Cuvier).

Identification: A coral snake with a red head and triads of black rings, of which the middle one is distinctly broader than the lateral ones. Adults average about 3 feet in length; occasional individuals attain a length of about 4 feet.

Crown of head red, with each of the plates outlined in black. Body with 5–8 complete triads, each made

FIGURE 36.—Surinam Coral Snake, *Micrurus surinamensis.* The red head and triad pattern are distinctive. Photo by Charles M. Bogert.

up of a broad middle black band, with narrow bands laterally. Yellow rings narrowed dorsally. Dorsals 17–19 anteriorly, 15 at midbody and posteriorly.

Ventrals 162–206; anal plate divided (occasionally entire); subcaudals 30–40.

Distribution: Apparently a semiaquatic snake (one specimen had eaten an eel) that inhabits the rim of the Amazon region; the Guianas, Brazil, Venezuela, Ecuador, Peru, and Bolivia.

Remarks: This is another of the large coral snakes. Its red head and the broad median band of the triad makes it distinctive.

CROTALIDAE: Genus *Bothrops* Wagler, 1824.

American lance-headed vipers.

Between 40 and 50 species are currently recognized; all are found in tropical America and southern South America. There are three general groups: 1. Large, long-tailed terrestrial species, usually with paired subcaudals: 2. Small, short-tailed terrestrial species with single subcaudals; and 3. Small to moderate-sized arboreal species with prehensile tail, most of which have at least the anterior subcaudals single. The large terrestrial species are very dangerous, the others less so.

Definition: Head broad, flattened, very distinct from narrow neck; a sharply-distinguished canthus. Body cylindrical or moderately compressed, moderately slender to stout; tail short to moderately long.

Eyes small to moderate in size; pupils vertically elliptical.

Head scales: Supraoculars generally present, internasals often distinct, sometimes separated by small scales; remainder of crown covered with small imbricate scales; enlarged canthals sometimes present. Laterally, second supralabial may make up anterior border of loreal pit or may be separated from it. Loreal scales present or absent.

Body scales: Dorsals keeled, in 19–35 nonoblique rows at midbody. Ventrals 121–253; subcaudals single or paired, 22–83.

Urutu, *Bothrops alternatus* (Duméril, Bibron, and Duméril).

Identification: A brown lancehead with rounded blotches which are narrowly edged with yellow. Adults average 3 to 4 feet; occasional individuals exceed 5 feet.

Head brown with a distinctive marking on the crown. About 20 pairs of rounded lateral markings shaped like a French telephone ⬤ whose apices nearly meet on the dorsal midline. Ground color brown, slightly lighter than blotches which have lighter centers. Belly white, spotted with brown or black.

Dorsals strongly keeled, in 29–35 rows at midbody. Ventrals 167–181; subcaudals paired, 34–51.

Distribution: Along watercourses through southern Brazil, Uruguay, Paraguay, and northern Argentina.

Remarks: This is a dangerous snake and it causes a large number of bites each year. Ordinarily the bite is not lethal, but it causes severe local effects.

FIGURE 37.—Urutu, *Bothrops alternatus.* Photo by New York Zoological Society. (See also plate II, fig. 2.)

A polyvalent antivenin "Antibotropico" is produced by the Instituto Butantan, and by the Instituto Pinhieros (Brazil).

Amazonian Tree Viper, *Bothrops bilineatus* (Wied).

Identification: A green tree viper with a yellow lateral stripe. Adults average 24 to 30 inches; maximum length about 3 feet.

Uniform bright green above, speckled with black in some individuals; a narrow yellow stripe or series of yellow spots on first row of dorsals. Tip of tail usually red or red-brown. Belly white, without markings.

Snout rounded; canthus rostralis sharp and slightly raised. Internasals large and in contact with one another; canthals large; 5–8 rows of scales between large supraoculars. Dorsals strongly keeled, in 27–35 nonoblique rows at midbody, fewer posteriorly. Ventrals 198–218; subcaudals 59–71, all or nearly all paired.

Distribution: The Amazonian regions of Brazil, British Guinea, Colombia, Bolivia, Peru, Ecuador, and Venezuela.

Remarks: This is one of the most widely distributed of the prehensile-tailed tree vipers of South America. However, it does not appear to be a serious hazard anywhere and no specific antivenin is produced for the treatment of its bite.

St. Lucia Serpent, *Bothrops caribbaeus* Garman.

Identification: A pale gray or yellowish gray pit viper; the only venomous snake on the West Indian island of St. Lucia. Adults average 3 to 4 feet in length; occasional individuals are recorded at about 7 feet.

Head dark gray with a postorbital band that extends across the upper edge of the supralabials. Body blotches obscure, little darker than the ground color which is light gray, often with rust-red suffusion. Chin white or cream, belly yellowish with a few gray markings.

Dorsals strongly keeled, in 25–29 rows at midbody, fewer (19) posteriorly. Ventrals 197–212; subcaudals paired, 64–70.

Distribution: Found in cacao and cocoanut plantations and damp forest; only on the island of St. Lucia.

FIGURE 38.—St. Lucia Serpent, *Bothrops caribbaeus.* Photo by New York Zoological Society.

Remarks: This is a dangerous snake whose bite causes severe local tissue damage.

For many years it was confused with the barba amarilla (*B. atrox*) of the mainland and the fer-de-lance (*B. lanceolatus*) of Martinique.

It is reported to have caused the death of several persons on the island. No specific antivenin is available for this species.

Jararaca, *Bothrops jajaraca* (Wied).

Identification: An olive-green, brown-blotched pit viper with a rather long, but short-snouted head. Adults average 3 to 4 feet; occasional individuals approach 6 feet in length.

Crown of head dark olive, usually with some dark brown irregular markings which may be light-edged. A

FIGURE 39.—Jararaca, *Bothrops jajaraca.* Photo by New York Zoological Society.

well-defined dark-brown postorbital stripe present; remainder of side of head light. About 25 pairs of lateral brown blotches on the body; they are well-defined lateral triangles anteriorly but become rounder toward midbody and quite irregular in shape posteriorly. Ground color olive, grayish or brownish. Belly yellowish, blotched with gray, often entirely gray posteriorly.

Prefrontals small, longer than broad, separated from one another by 4–5 rows of small scales. Dorsals weakly keeled, keels extending entire length of scales, in 20–27 rows at midbody. Ventrals 175–216; subcaudals 52–70, all or nearly all paired.

Distribution: Grasslands and open country through southern Brazil, northeastern Paraguay and northern Argentina.

Remarks: This snake is easily confused with *B. atrox* on the one hand and with *B. jararacussu* on the other. The color patterns and scales of the snout region appear to distinguish them. *B. jajaraca* is one of the most common venomous snakes throughout its range. Probably for that reason, rather than because of its venom quantity and toxicity, it is second only to the cascabel (*Crotalus durissus*) as a source of deaths from snakebite in the region.

Jararacussú, *Bothrops jararacussu* Lacerda.

Identification: A dull-colored black and yellowish pit

FIGURE 40.—Jararacussú, *Bothrops jararacussu.* Photo by New York Zoological Society.

viper with a broad, lance-shaped head. Adults average 3 to 4 feet; maximum length about 5½ feet (Amaral, 1925).

Crown of head unicolor black and dark brown with dark-yellowish lines over the temporal regions which separate the black postorbital stripe from the dark color of the crown; side of head mostly yellowish. About 15 pairs of lateral upside-down U-shaped black body blotches may alternate with one another or oppose and connect across the back. Often much of back covered with irregular patches of dark pigment., leaving lateral blotches irregularly outlined with dark yellow. Belly yellow, irregularly blotched with dark brown or black.

Prefrontals (canthals) broader than long, separated from one another by 1–2 rows of small scales. Dorsals strongly keeled, keels tending to be tuberculate along back, in 23–27 rows at midbody, fewer posteriorly. Ventrals 170–186; subcaudals 44–66, all or nearly all paired.

Distribution: Near rivers and lakes in southern Brazil, eastern Bolivia, Paraguay, and northern Argentina.

Remarks: This is an amphibious species and may be found in the water. It is not a very common snake, but produces a very toxic venom in large amounts (averaging more than 100 mg. in a milking); it is one of four species of snakes which cause most fatalities from snakebite in Brazil. A common early symptom of its bite is blindness.

Antivenins (polyvalent) using the venom of *B. jararacussu* are produced by Behringwerke of Germany, Instituto Butantan and Instituto Pinheiros of Brazil, and the Instituto Nacional de Microbiologia of Argentina.

Fer-de-Lance, *Bothrops lanceolatus* (Lacépède).

Identification: A lancehead recognized by its dark truncated lateral blotches and high numbers of dorsals and ventrals; the only venomous snake on Martinique. Adults average 4 to 5 feet; occasional individuals attain lengths of about 7 feet.

Head brown with a sharply defined darker postorbital band that extends down to the corner of the mouth. Body gray, olive, or brown with an obscure series of 22–27 hour-glass-shaped blotches down the back. Ventral surface white or cream with a few grayish or brown stipple marks anteriorly, more posteriorly.

Dorsals strongly keeled, in 31–33 rows at midbody, fewer (29) anteriorly and posteriorly (21–23). Ventrals 215–230; subcaudals paired, 56–67.

Figure 41.—Fer-de-Lance, *Bothrops lanceolatus*. The snake to which the name, Fer-de-Lance, rightfully belongs is found only on the island of Martinique. Photo by New York Zoological Society.

Distribution: Found only on the West Indian island of Martinique; originally over the entire island but now restricted to the less inhabited forests.

Jararaca pintada, *Bothrops neuwiedi* Wagler.

Identification: A distinctly-patterned tan or grayish pit viper with a distinctive pattern on the crown. Adults average 2 to 3 feet in length.

Crown of head light tan or brown with a series of distinct spots; often a U-shaped mark on the rear part of the head, the two arms of the "U" sometimes connected with the body pattern. Pattern geographically variable but basically a paired series of small triangular or rhomboidal black or dark brown dorsal blotches that alternate or fuse across the back to form small X-shaped markings. Rounded dark spots may fall between the main series on the midline and a lateral series of small spots alternates with the dorsal blotches. All of the

Figure 42.—Jararaca pintada, *Bothrops neuwiedi*. The "U" mark on the rear of the head is distinctive. Photo by Isabelle Hunt Conant.

markings may be outlined with bright yellow. Ground color tan or light gray. Belly yellowish, some ventrals edged with gray.

Dorsals strongly keeled, in 21–27 rows at midbody. Ventrals 163–187; subcaudals 40–53, all paired.

Distribution: Grasslands and open country on the plateau of southern Brazil, eastern Bolivia, Paraguay, and northern Argentina.

Remarks: This is a rather small snake but it ranges over a large area of southern South America. It is one of the major sources of snakebite in Argentina.

Polyvalent antivenins are produced by Behringwerke of Germany, and the Instituto Nacional de Microbiologia, Argentina.

CROTALIDAE: Genus *Crotalus* Linnaeus, 1758.

Rattlesnakes.

About 25 species of rattlesnakes are currently recognized. Most species are in the southwestern United

States and northern Mexico; one species (*C. durissus*) ranges southward into southern South America, two are found east of the Mississippi River, and two as far north as Canada. A few of the very small species and small individuals of large species (less than 2 feet) may offer little danger, but most species do; some are highly dangerous.

Definition: Head broad, very distinct from narrow neck; canthus distinct to absent. Body cylindrical, depressed, or slightly compressed, moderately slender to stout; tail short with a horny segmented rattle.

Eyes small; pupils vertically elliptical.

Head scales: Supraoculars present, a pair of internasals often distinct, occasionally a pair of prefrontals; enlarged canthal scales often present; other parts of crown covered with small scales. Laterally, eye separated from supralabials by 1–5 rows of small scales.

Body scales: Dorsals keeled, with apical pits, in 19–33 nonoblique rows at midbody. Ventrals 132–206; subcaudals 13–45, all single or with some terminal ones paired.

Cascabel, *Crotalus durissus* Linnaeus.

Identification: The only true rattlesnake in most of its range (except in Mexico). The series of large rhombic blotches (diamonds) down the back, stripes on the neck, and the large rattle are distinctive. Body stout and slightly compressed, especially anteriorly. Adults average 4 to 5 feet; maximum length about 6 feet.

Body brown or olive with 18–35 darker, light-edged rhomb-shaped markings down the back. Those on neck sometimes elongate into stripes. Tail usually unicolor dark brown or black. White or cream colored below.

FIGURE 43.—Cascabel, *Crotalus durissus*. Photo courtesy *Scientific American*. (See also plate I, fig. 3.)

Distribution: Dry areas, grasslands, and thorny scrub, from coastal eastern and southern Mexico southward through Central America, and through eastern South America from northern Colombia to northern Argentina.

Remarks: This is one of the most dangerous of the rattlesnakes, and is one of the most dangerous snakes in the Americas. The toxicity of the venom varies through the range; in Brazil, where the cascabel is the main cause of death from snakebite, it is extremely toxic. The venom of this rattlesnake has minor local effect but very grave systemic symptoms. These include blindness, paralysis of the neck muscles, cessation of breathing and heartbeat, and finally death.

This venom does not appear to form adequate antibodies in horses, so that enormous amounts of antivenin are needed to counteract the effects of the bite of a snake of average dimensions. Ten ampules (100 ml.) would appear to be an average initial dose, and 20 or more may be used.

Antivenins are produced by the Instituto Butantan and Instituto Pinheiros, Brazil, and Wyeth, Inc., Philadelphia.

Aruba Rattlesnake, *Crotalus unicolor* Lidth de Jeude.

Identification: A gray or gray-brown rattlesnake which is unicolor, or with a faint pattern of rhomb-

FIGURE 44.—Aruba Rattlesnake, *Crotalus unicolor*. This faded relative of the cascabel occurs only on the island of Aruba. Photo by New York Zoological Society.

shaped blotches (diamonds) down the back; the only venomous snake on Aruba Island. Body stout and somewhat depressed. Adults average 2 to 3 feet; maximum length a little less than 38 inches (950 mm.; Klauber, 1956).

Body gray or light gray-brown with 18–28 faintly darker rhomb-shaped blotches down the back; blotches sometimes almost indistinguishable. A lateral series of obsolete blotches that alternate with or oppose the dorsal series. Usually a distinct pair of parallel stripes on the rear part of the head; these may continue as stripes on the neck. White or cream-colored below.

Dorsals strongly keeled, in 25–27 rows at midbody,

fewer posteriorly. Ventrals 155–169; subcaudals 22–31.

Distribution: Found only on the island of Aruba, in the Caribbean Sea, off the coast of Venezuela.

Remarks: This is a dwarfed and light-colored relative of the cascabel (*Crotalus durissus*). It is not aggressive but ready to defend itself. Nothing is known of its venom but the close relationship with the cascabel suggests that it is capable of a dangerous bite in spite of its small size.

CROTALIDAE: Genus *Lachesis* Daudin, 1803.

Bushmaster.

A single species, *L. mutus*, is found in tropical America. It attains a length of 9 to 12 feet and is considered dangerous (see pp. 56–57).

Definition: Head broad, very distinct from narrow neck; snout broadly rounded, no canthus. Body cylindrical, tapered, moderately stout; tail short.

Eyes small; pupils vertically elliptical.

Head scales: A pair of small internasals separated from one another by small scales; a pair of narrow supraoculars; other parts of crown covered with very small scales. Laterally, second supralabial forms anterior border of loreal pit, third very large; eye separated from supralabials by 4–5 rows of small scales.

Body scales: Dorsals heavily keeled with bulbous tubercles, feebly imbricate, in 31–37 nonoblique rows at midbody, fewer posteriorly. Ventrals 200–230; subcaudals mainly paired, 32–50, followed by 13–17 rows of small spines and a terminal spine (fig. 32).

REFERENCES
(See also General References)

AMARAL, Afranio do. 1925. A General Consideration of Snake Poisoning and Observations on Neotropical Pit-Vipers. Contr. Harvard Inst. Tropical Biol. Med., 2: 64 p., 16 pls.

AMARAL, Afranio do. 1930. Lista remissiva dos ophidios da regiao neotropica. Mem. Inst. Butantan, 4: 127–271.

DUNN, Emmett R. 1944. Los genereos de anfibios y reptiles de Colombia, III. Tercera Parte: Reptiles; orden de las serpientes. Caldasia, 3 (12): 155–224.

FONSECA, Flavio da. 1949. Animais peconhentos. Publ. Inst. Butantan, Sao Paulo. 376 p., 129 figs., 13 color pls.

HOGE, Alphonse R. 1965. Preliminary Account on Neotropical Crotalinae (Serpentes, Viperidae). Mem. Inst. Butantan 32:109–184, pls. 1–20, maps 1–10.

LAZELL, James D., Jr. 1964. The Lesser Antillean Representatives of *Bothrops* and *Constrictor*. Bull. Mus. Comp. Zool., 132 (3): 245–273, figs. 1–5.

MOLE, R. R. 1924. The Trinidad Snakes. Proc. Zool. Soc. London: 235–278, pls. 1–10.

PETERS, James A. 1960. The Snakes of Ecuador: A Check List and Key. Bull. Mus. Comp. Zool., 122 (9): 491–541.

ROZE, Janis A. 1955. Revision de las corales (Serpentes: Elapidae) de Venezuela. Acta Biol. Venezuelica, 1 (art. 17): 453–500, 4 figs. (2 color).

ROZE, Janis A. 1966. La Taxonomia y Zoogeografia de los ofidios de Venezuela. Universidad Central de Venezuela, Caracas. 362 pp. 79 figs., 80 maps.

SANDNER MONTILLA, F. 1965. Manual de las serpientes ponzonosas de Venezuela. (Pub. by author) Caracas. 108 p. 69 figs., 9 col. pls.

SANTOS, Eurico. 1955. Anfibios e repteis do Brasil (vida e costumes). 2nd ed. F. Briguiet & Cia., Rio de Janeiro. 262 p., 65 figs., 10 color pls.

NOTES

Section 4

EUROPE

Definition of the Region:

Entire continent of Europe, European Russia (Russian Soviet Federated Socialist Republic) and the Mediterranean islands, the Ukranian SSR and the Autonomous Soviet Republics north of the Caucasus and west of the Volga River.

TABLE OF CONTENTS

TABLE 8.—DISTRIBUTION OF POISONOUS SNAKES IN EUROPE*

	British Isles (except Ireland)	Scandanavia & Finland	Poland	Germany, Netherlands, Belgium	Czechoslovakia	Austria & Switzerland	Hungary	France	Spain & Portugal	Italy	Yugoslavia & Albania	Greece (Mainland)	Greece (Cyclades & Cyprus)	Bulgaria & European Turkey	Romania	European Russia
VIPERIDAE																
Vipera ammodytes						SE Austria				NE	X	N	X	X	X	S
Vipera aspis				S Germany		X	SW	S	N	X	N					
Vipera berus	X	X	X	X	X	X	X	X	NW	N	X			W	W	X
Vipera kaznakovi																SW
Vipera latasti									X							
Vipera lebetina													X			S
Vipera ursinii					?	SE Austria	X	SE		X	X			X	X	S
Vipera xanthina														Turkey		S
CROTALIDAE																
Agkistrodon halys																E

*Certain groups of adjoining nations or provinces are here treated as units. The symbol X indicates distribution of the species is widespread within the unit. Restriction of a species to part of a unit is indicated appropriately (SW = southwest, etc.).

INTRODUCTION

Europe has comparatively few species of native snakes. This reflects the generally cool, present-day climate, the scarcity of suitable habitats for snakes, and the geologic history of glaciation that eliminated all reptiles from much of the continent some 10,000 to 20,000 years ago. Poisonous snakes in Europe tend to be quite local and spotty in distribution, especially toward the north. The hardwood and evergreen forests that originally covered much of the continent were never good habitats for snakes. Centuries of intensive agriculture and more recent industrialization have further reduced the suitable habitats. In spite of this, poisonous snakes may be locally plentiful. In Scandinavia and Finland, the European viper ranges slightly above the Arctic Circle—farther north than any other known species of snake. In Finland during the summer of 1961, 163 snakebites were reported. One physician in Cornwall, England, saw 18 cases of adder bite between 1952 and 1959. The eastern Mediterranean region has the greatest number of venomous snakes and the most dangerous species.

All the European poisonous snakes are vipers and present a strikingly similar appearance. They are small to medium-sized snakes of moderately stout build with short tails. In distinguishing them from nonpoisonous snakes, note that the eye is separated from the upper lip shields by one or more small scales (except in the single species of pit viper *Agkistrodon halys*) and the pupil is elliptical. In most European nonpoisonous snakes the eye touches the upper lip shields and the pupil is round. The only exceptions to both these rules are the little boas of the genus *Eryx;* they are easily recognized by their small ventrals. In distinguishing one species of

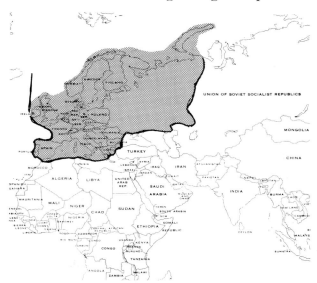

MAP 5.—Section 4, Europe.

viper from another, note particularly the shape of the snout and the presence or absence of enlarged shields on the top of the head. Body scales are keeled in all the European vipers.

The common vipers of Europe feed largely upon lizards and small mammals. They are all live-bearing.

Antivenins against venoms of the common vipers of Europe are produced by the Institut Pasteur, Paris; Behringwerke, Marburg-Lahn, Germany; Instituto Sieroterapico e Vaccinogeno Toscano, Siena, Italy; and the Institute for Immunology, Zagreb, Yugoslavia (*Vipera ammodytes* only).

KEY TO GENERA

1. A. Nine large crown shields (fig. 6); eye in contact with upper lip shields_____ 2
   B. Crown shields 6 or fewer or broken up into small scales; eye separated from lip shields_____ 3
2. A. Loreal pit present (fig. 4)_____ *Agkistrodon*
   B. Loreal pit absent_____ NP*
3. A. Ventrals extend full width of belly (fig. 9A)_____ *Vipera*
   B. Ventrals do not extend full width of belly (fig. 9B)_____ NP

_____
* NP = Nonpoisonous

GENERIC AND SPECIES DESCRIPTIONS

VIPERIDAE: Genus *Vipera* Laurenti, 1768.

True adders.

Eleven species are recognized. This is an especially variable group, with some members that are small and relatively innocuous (e.g., *V. berus*) and others that are extremely dangerous (*V. lebetina*, *V. russelii*). They are found from northern Eurasia throughout that continent and into north Africa. One species ranges into the East Indies (*V. russelii*), and two are found in east Africa (see Remarks under *V superciliaris*).

Definition: Head broad, distinct from narrow neck; canthus distinct. Body cylindrical, varying from moderately slender to stout; tail short.

Eyes moderate in size to small; pupils vertically elliptical.

Head scales: Variable: one species (*V. ursinii*) has all 9 crown scutes, most species have at least the supraoculars, but even these are absent in one (*V. lebetina*): head otherwise covered with small scales. Laterally, nasal in contact with rostral or separated by a single enlarged scale (the nasorostral), eye separated from supralabials by 1–4 rows of small scales.

Body scales: Dorsals keeled, with apical pits, in 19–31 nonoblique rows at midbody. Ventrals rounded, 120–180; subcaudals paired, 20–64.

European Viper, *Vipera berus* (Linnaeus).

Identification: Head distinct from neck but ovoid rather than distinctly triangular; snout blunt, flat, not upturned; top of head with 5 large smooth shields.

Ground color pale gray, olive or yellow to russet or brown, the darker colors generally in females. Down the entire length of the back runs a black or dark brown zigzag line rarely broken into spots for all or part of its length and even more rarely straight edged. Top of head behind eyes with a dark "X"—or chevron —mark; belly pale gray with darker suffusion. Uniformly black or very dark brown individuals are seen especially in some mountainous regions.

Average length 19 to 24 inches, maximum 34 inches; females larger than males.

Distribution: The only poisonous snake of northern Europe where it is widely distributed; in central and southern Europe largely confined to mountains where it occurs to at least 9,000 feet elevation. It ranges completely across northern Asia to the Russian island of Sakhalin and northern Korea. In the north usually found in dry open sunny places—moors, old fields, brushy hillsides and openings in the forest. In the south more prevalent on rocky hillsides and about the edges of mountain forests.

Remarks: Nocturnal during warm weather; diurnal in cool; has considerable tolerance for cold and may be seen basking near patches of snow. Disposition generally timid, but strikes quickly and repeatedly when cornered or suddenly alarmed.

Venom yield small but venom of fairly high toxicity.

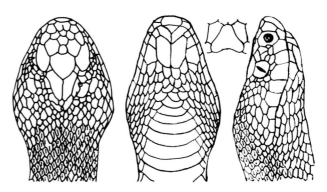

FIGURE 45.—Head scales of European Viper, *Vipera berus*. The broken-up crown shields on the snout are characteristic of this species. (See also plate II, fig. 1.) Redrawn from Maki, 1931.

Asp Viper, *Vipera aspis* (Linnaeus).

Identification: Head more triangular than in European viper, snout slightly but distinctly upturned at tip; shields on crown fragmented, usually only 2 or 3 enlarged.

Color similar to European viper but generally more apt to be reddish or brown; pattern of dark spots more or less fused, sometimes forming zigzag band; dark head mark not well defined; belly dark gray with lighter flecks; underside of tail tip yellow or orange.

FIGURE 46.—Asp Viper, *Vipera aspis*. Photo by Isabelle Hunt Conant.

Size about the same as European viper, 18 to 24 inches; males average larger than females.

Distribution: The western part of southern Europe. Found mostly in hilly or mountainous country to an altitude of 7,800 feet in the Pyrenees.

Remarks: Disposition generally more sluggish than European viper. Venom of about the same toxicity.

Snub-nosed Viper, *Vipera latasti* Boscá.

Identification: Similar to the asp viper but snout more upturned and pointed, its anterior surface formed only from the rostral; shields of crown much fragmented and usually not symmetrical.

Color as in the other two species; zigzag dorsal line prominent and well defined.

Size about the same as the European viper.

Distribution: The Iberian Peninsula and northwest Africa. Found in lowlands and at moderate elevations usually in open sandy or rocky terrain.

Remarks: Little known of the venom, but it is not believed to be a particularly dangerous species.

Long-nosed Viper, *Vipera ammodytes* (Linnaeus).

Identification: Most readily identified by the snout which terminates in a strongly upturned appendage, its anterior surface formed from several small scales; crown

FIGURE 47.—Long-nosed Viper, *Vipera ammodytes.* Photo by New York Zoological Society.

covered by small scales of irregular size and arrangement.

Color ash-gray, yellow, pale orange, coppery or brownish; zig-zag dorsal line very prominent; pattern more vivid in male; head without distinct dorsal markings; belly yellow or brownish more or less heavily clouded with dark gray; tail tip orange or reddish.

Average length 25 to 30 inches; maximum about 36 inches. Males are larger than females.

Distribution: Southeastern Europe and Asia Minor. Inhabits dry hilly country for the most part between 2,000 and 5,500 feet elevation. It prefers rocky slopes particularly where there are outcrops of limestone.

Remarks: Largely nocturnal but may be active by day in cool weather. Sometimes climbs onto bushes to bask in the sun. Rather sedentary and retiring in habits but quick to strike. It is generally thought to be the most dangerous of the European vipers. The venom is quite toxic and apparently varies considerably in composition over the range of the species.

Two large vipers just enter European territory, the Ottoman viper (*Vipera xanthina*) near Istanbul and the Levantine viper (*Vipera lebetina*) on some of the eastern Mediterranean islands. (For descriptions of these species, see page 111 and page 112.)

CROTALIDAE: Genus *Agkistrodon* Beauvois, 1799.

Moccasins and Asian pit vipers.

Twelve species are recognized. Three of these are in North and Central America; the others are in Asia, with one species, *A. halys* (Pallas) ranging westward to southeastern Europe. The American copperhead (*A. contortrix*) and the Eurasian mamushi and its relatives (*A. halys*) seldom inflict a serious bite, but *A. acutus* and *A. rhodostoma* of southeastern Asia, as well as the cottonmouth (*A. piscivorus*) of the southeastern United States are dangerous species.

Definition: Head broad, flattened, very distinct from narrow neck; a sharply-distinguished canthus. Body cylindrical or depressed, tapered, moderately stout to stout; tail short to moderately long.

Eyes moderate in size; pupils vertically elliptical.

Head scales: The usual 9 on the crown in most species; internasals and prefrontals broken up into small scales in some Asian forms; a pointed nasal appendage in some. Laterally, loreal pit separated from labials or its anterior border formed by second supralabial. Loreal scale present or absent.

Body scales: Dorsals smooth (in *A. rhodostoma* only) or keeled, with apical pits, in 17–27 nonoblique rows. Ventrals 125–174; subcaudals single anteriorly or paired throughout, 21–68.

Pallas' Viper, *Agkistrodon halys* (Pallas).

Identification: The loreal pit distinguishes this species from all other snakes of Europe and central Asia. Presence of 9 large head shields and contact of at least one supralabial with the eye, distinguish it from other vipers of that region. The pit and generally viperine

FIGURE 48.—Pallas' Viper, *Agkistrodon halys intermedius.* Specimen from Uzbek, U. S. S. R. Photo by Sherman A. Minton. (Preserved specimen)

body form distinguish it from the comparatively few nonpoisonous snakes within its range.

Color yellowish, tan or grayish with many dark brown or gray crossbands alternating with spots on the sides or with crossbands and spots fusing to produce an irregular network; belly cream to yellow with fine black punctation especially toward the tail; top of head with dark spot above each eye and at nape; tip of tail yellowish. Average length 22 to 28 inches; maximum about 35 inches.

Distribution: A characteristic snake of the vast central Asian steppe where it occurs in grassland and desert; often abundant around rocky bluffs that probably are hibernating dens. Range in Europe restricted to the region between the Volga and the Urals; found eastward to southern Siberia and Mongolia.

Remarks: Largely nocturnal; rests during day beneath stones or shrubs. Bites by this snake are not infrequent, but fatalities are rare.

This account deals chiefly with *Agkistrodon h. halys, A.h. caraganus,* and *A.h. intermedius.* The races of *A. halys* in the Far East are treated elsewhere.

REFERENCES

HELLMICH, Walter. 1956. Die Lurche und Kriechtiere Europas. Carl Winter: Heidelberg, pp. 1–166, pls. 1–68 (color), figs. 1–9.

MERTENS, Robert and H. WERMUTH. 1960. Die Amphibien und Reptilien Europas. Dritte Liste nach dem Stand vom 1. Januar 1960. Waldemar Kramer: Frankfurt-am-Main, xi + 264 pp., figs. 1–45.

SCHWARZ, E. 1936. Untersuchungen uber Systematik und Verbreitung der europaischen und mediterranen Ottern. Behringwerk-Mitteilung: Marburg-Lahn, vol. 7, pp. 159–362, pl. 1.

TERENTJEV, P. V. and S. A. CERNOV. 1949. Apredelitelj Presmykajuschcichsja i Semnowodnych. [Distribution of Reptiles and Amphibians] Third Ed. [in Russian] Government Printing Office: Moscow, 339 pp., 123 figs., 37 maps.

NOTES

Section 5

NORTH AFRICA

Definition of the Region:
Includes the nations of Mauritania, Spanish Sahara, Mali, Niger, Morocco, Algeria, Tunisia, Libya, and Egypt (United Arab Republic).

TABLE OF CONTENTS

TABLE 9.—DISTRIBUTION OF POISONOUS SNAKES IN NORTH AFRICA

	United Arab Republic (Egypt)	Libya	Algeria	Tunisia	Morocco—IFNI	Spanish Sahara	Mauritania	Mali	Niger	Chad
ELAPIDAE										
Elapsoidea sundevallii							S	S		
Naja haje	X	X	N	X	S	X	S	X	X	?
Naja nigricollis	S						S	S	S	?
Walterinnesia aegyptia	NE									
VIPERIDAE										
Atractaspis engaddensis	NE									
Atractaspis microlepidota							S	S	S	?
Bitis arietans			W		S	X	S	X	X	?
Causus rhombeatus							S	S	S	?
Cerastes cerastes	X	X	X	S	X	X	X	N	N	?
Cerastes vipera	X	X	X	S	X	X	X	N	N	
Echis carinatus	X	X	X	S	SW	X	X	X	X	?
Echis coloratus	E									
Vipera latasti				N	X					
Vipera mauritanica		N		X	X	N				N

The symbol X indicates distribution of the species is widespread within the country. Restriction of a species to part of a country is indicated appropriately (SW = southwest, etc.). The symbol ? indicates suspected occurrence of a species in the unit without valid literature.

INTRODUCTION

Africa from the southern edge of the Sahara northward is a vast region where the dominant theme is heat and aridity. This is mitigated only along the Mediterranean coast, in high mountains such as the Atlas range, and along the great river valleys and oases.

The snake fauna contains few species partly because of the rigors of the climate, and partly because most of the desert is new and there has been insufficient time for the evolution and spread of a specialized desert snake fauna. The distribution of snake species in northern Africa is not well known. There are probably a number of tropical African species that invade locally along the large rivers in the southern part of the region. Only a few of the species are found primarily in the desert; the majority occur around zones of irrigation or natural water supply. This increases the hazard of snakebite to the rural

people; however, the incidence of such accidents is unknown. Egypt in the years 1944–48 reported

MAP 6.—Section 5, North Africa.

KEY TO GENERA

1. A. Crown of head covered with small irregular scales;
      pupil of eye vertically elliptical_____ 7
   B. Crown of head covered with large shields; pupil
      round or elliptical_____ 2
2. A. Loreal plate present_____ 3
   B. Loreal plate absent_____ 4
3. A. Lateral scales rectangular and oblique; top of head
      with dark chevron marking (plate VIII, fig. 4) *Causus rhombeatus*
   B. Without the above combination of characters_____ NP*
4. A. Eye very small, snout pointed, all subcaudals
      undivided_____ *Atractaspis*
   B. Without the above combination of characters_____ 5
5. A. All dorsal scales smooth_____ 6
   B. Posterior dorsal scales keeled; anal plate divided_____ *Walterinnesia*
6. A. Scale rows at midbody more than 15; hood seen
      in life_____ *Naja*
   B. Scale rows at midbody usually 13; no hood_____ *Elapsoidea*
7. A. Lateral scales oblique with serrated keels_____ 8
   B. Lateral scales like dorsals_____ 9
8. A. Subcaudals single; ventrals not keeled_____ *Echis*
   B. Subcaudals paired; ventrals keeled_____ *Cerastes*
9. A. Ventrals extending full width of belly_____ 10
   B. Ventrals not extending full width of belly_____ NP
10. A. Body pattern of chevron-shaped crossbands;
       nostrils dorsal_____ *Bitis*
    B. Body pattern not as above; nostrils lateral_____ *Vipera*

_____
* NP = Nonpoisonous

26 to 46 snakebite deaths annually; the true figure is probably higher.

The most important poisonous snakes of north Africa are vipers; cobras occur but apparently play a minor role in snakebite accidents.

GENERIC AND SPECIES DESCRIPTIONS

ELAPIDAE: Genus *Elapsoidea* Bocage, 1866.

African garter snakes.

A single species (*E. sundevallii*) with 11 geographic races is currently recognized (See p. 94). It ranges over most of tropical and southern Africa except for the Cape region. It attains a length of 3 to 4 feet and is potentially dangerous. However, it is sluggish and inoffensive and bites only in self-defense. This species enters the southern part of this region (see plate VIII, fig. 3).

Definition: Head of moderate size, not distinct from neck; an indistinct canthus. Body moderately slender, cylindrical; tail very short.

Eyes small; pupils round.

Head scales: The usual 9 on the crown; rostral enlarged, obtusely pointed; internasals short. Laterally, nasal in narrow contact with single preocular.

Body scales: Dorsals smooth and rounded, in 13 rows at midbody. Ventrals 138–184; anal plate entire; subcaudals paired (a few sometimes single) 13–29.

Maxillary teeth: Two large tubular fangs with external groove followed, after an interspace, by 2–4 small teeth.

ELAPIDAE: Genus *Naja* Laurenti, 1768.

Cobras.

Six species are recognized; all are African except the Asiatic cobra, *Naja naja*, and range throughout the African continent except for the drifting sand areas of the Sahara region. They are snakes of moderate (4 feet) to large (8 feet) size, with large fangs and toxic venom. The species *N. nigricollis* "spits" its venom at the eyes of an aggressor; it is found in the southern part of the region of north Africa. The Egyptian cobra (*Naja haje*) and the western subspecies of the Asiatic cobra (*Naja naja oxiana*) are found in the Near and Middle East region.

Definition: Head rather broad, flattened, only slightly distinct from neck; snout rounded, a distinct canthus. Body moderately slender, slightly depressed, tapered, neck capable of expansion into hood; tail of moderate length.

Eyes moderate in size; pupils round.

Head scales: The usual 9 on the crown; frontal short; rostral rounded. Laterally, nasal in contact with the one or two preoculars.

Body scales: Dorsals smooth, in 17–25 oblique rows at midbody, usually more on the neck, fewer posteriorly. Ventrals 159–232; anal plate entire; subcaudals 42–88, mostly paired.

Maxillary teeth: Two rather large tubular fangs with external grooves followed, after an interspace, by 0–3 small teeth.

Egyptian Cobra, *Naja haje* (Linnaeus).

Identification: Body form typically cobra like—short wide head, not distinct from neck; body moderately stout but graceful with even taper and moderately long tail; scales smooth with dull sheen, scale rows strongly oblique especially on forebody; anal plate entire; subcaudals paired.

A useful point in identification of cobras and cobra-like venomous snakes (elapids) is the absence of the loreal shield so that the shield bordering or enclosing the nostril touches the shield that borders the eye anteriorly (the preocular). The loreal is present in most nonpoisonous snakes, and absent most often in small burrowing or secretive types. The Egyptian cobra may be distinguished from other African cobras by the presence of small subocular scales separating the eye from the upper labials.

Color extremely variable. Adult snakes from Egypt and Libya may be brownish yellow, dark brown, or almost black; the head and neck are almost always a little darker; below yellowish becoming suffused with

FIGURE 49.—Egyptian Cobra, *Naja haje.* Photo by Isabelle Hunt Conant. (See also plate VIII, fig. 7.)

brown; dark bars across neck at level of hood. Young yellowish; head and neck black; body crossed by wide dark bands. Adult snakes from southern Morocco are black above; purplish red with black bars and mottling below.

A large cobra, maximum length about 8 feet; average 5 to 6 feet.

Distribution: Occurs throughout the northern three-quarters of Africa exclusive of the rain forest; also found in the western and southern parts of the Arabian Peninsula.

Found in a great variety of habitats such as flat land with scrubby bushes and grass clumps; irrigated fields, rocky hillsides, old ruins and in the vicinity of villages.

It avoids extreme desert situations and also permanently moist ones. Like many snakes it often makes its home in abandoned rodent burrows.

Remarks: While there are reports of aggressive behavior by Egyptian cobras, this is exceptional. They seem to be rather timid snakes and often make little effort to defend themselves. The hood is not so wide as in the Indian cobra.

The cobra type of defense with the body raised high off the ground and neck spread is impressive and helpful in recognition of these snakes when they are alive. It is important to remember, however, that cobras may bite without spreading the hood and occasionally may spread the hood without rearing up the forebody. It should also be noted that many unrelated nonpoisonous snakes in various parts of the world spread the neck and forebody.

The venom is of about the same degree of toxicity as that of the Indian cobra. If Cleopatra really used one of Egypt's snakes as an instrument of suicide, this species would have been a wise choice. Antivenin against venom of this cobra is produced by the Institut Pasteur, Paris, and Behringwerke, Marburg-Lahn, Germany.

This cobra is the sacred snake (Uraeus) of ancient Egypt and is probably the snake known as asp to the classical writers of Greece and Rome.

ELAPIDAE: Genus *Walterinnesia* Lataste, 1887.

Desert black snake.

A single species, *W. aegyptia*, is known from the desert regions of Egypt to Iran. It is relatively large, 3 to 4 feet, and is probably a dangerous species.

Definition: Head relatively broad, flattened, distinct from neck; snout broad, a distinct canthus. Body cylindrical and tapered, moderately slender; tail short.

Eyes moderate in size; pupils round.

Head scales: The usual 9 on the crown; rostral broad. Laterally, nasal in contact with single elongate preocular.

Body scales: Dorsals smooth at midbody, feebly keeled posteriorly, in 23 rows at midbody, more (27) anteriorly. Ventrals 189–197; anal plate divided; subcaudals 45–48, first 2–8 single, remainder paired.

Maxillary teeth: Two large tubular fangs with external grooves followed, after an interspace, by 0–2 small teeth.

Desert Black Snake, *Walterinnesia aegyptia* Lataste.

Identification: A moderately stout snake with short tail and small head not distinct from neck; crown with large shields. The following combination of scale characters is useful in distinguishing this species from nonpoisonous snakes and cobras: 1. Loreal plate absent; 2. Dorsal scales smooth anteriorly, keeled posteriorly; 3. Anal plate divided; 4. Some single subcaudals, although most are paired.

Adults uniformly black or very dark brown or gray

above, a little paler ventrally. Young, in Iran at least, have narrow light crossbands.

Average length 3 to 3½ feet; maximum a little over 4 feet.

Distribution: Egypt and the nations of the Near and Middle East. Reported most frequently from gardens, oases and irrigated areas; also inhabits barren rocky mountain hillsides and sandy desert with sparse bushes. A rather rare snake.

Remarks: Does not rear up or spread hood but when annoyed may strike more than half its length. The high gloss of its scales helps to distinguish this species from the duller Egyptian cobra.

Toxicity of the venom for experimental animals is about the same as that of the Indian cobra but quantity is considerably less (about 20 mg. vs. 50 to 100 mg.). There is no antivenin available.

FIGURE 50.—Desert Black Snake, *Walterinnesia aegyptia.* The highly glossy scales help to differentiate this snake from other dark species within its range. Photo courtesy Standard Oil Company.

VIPERIDAE: Genus *Atractaspis* Smith, 1949.

Mole vipers.

Sixteen species are currently recognized. All are African except for *A. engaddensis* Haas (which ranges from Egypt to Israel) and *A. microlepidota* (which is found in the southern part of the Arabian Peninsula as well as through much of northern and central Africa). All are small snakes, less than 3 feet in length. However, they have large fangs (which look enormous in their small mouths) and are capable of inflicting serious bites to those picking them up or stepping on them with bare feet (see p. 99).

Definition: Head short and conical, not distinct from neck, no canthus; snout broad, flattened, often pointed. Body cylindrical, slender in small individuals, stout in large ones; tail short, ending in a distinct spine.

Eyes very small; pupils round.

Head scales: The usual 9 crown scales, rostral enlarged, extending between internasals to some degree, often pointed; frontal large and broad, supraoculars small. Laterally, nasal in contact with single preocular (no loreal), usually one postocular.

Body scales: Dorsals smooth without apical pits, in 19–37 nonoblique rows at midbody. Ventrals 178–370; anal plate entire or divided (the only viperid snake with divided anal plates); subcaudals single or paired, 18–39.

VIPERIDAE: Genus *Bitis Gray,* 1842.

African vipers.

Ten species are found in tropical and southern Africa. They include the largest of the true vipers (Viperidae) as well as some small and moderately sized ones; all of the members of the genus are dangerous, some of them excremly so. The puff adder, *Bitis arictans,* is found widely through the region (see p. 101).

Definition: Head broad and very distinct from narrow neck; snout short, a distinct canthus. Body somewhat depressed, moderately to extremely stout; tail short.

Eyes small; pupils vertically elliptical.

Head scales: No enlarged plates on crown, covered with small scales. Some species have enlarged and erect scales on snout or above eye. Laterally, rostral separated from nasal by 0 (in *B. worthingtoni*) to 6 (in some *B. nasicornis*) rows of small scales, eye separated from supralabials by 2–5 rows of small scales.

Body scales: Dorsals keeled with apical pits, in 21–46 nonoblique or slightly oblique rows at midbody, fewer anteriorly and posteriorly. Ventrals rounded or with faint lateral keels, 112–153; subcaudals paired, laterally keeled in some species, 16–37.

VIPERIDAE: Genus *Causus Wagler,* 1830.

Night adders.

Four species are found in tropical and southern Africa. None attains a length of over 3 feet. The fangs are relatively small, and the venom is rather mildly toxic. They look surprisingly like nonpoisonous snakes. Night adders are not considered dangerous to life but their bite is painful and venomous. The rhombic night adder, *C. rhombeatus,* enters the southern part of this region (p. 102).

Definition: Head moderate in size, fairly distinct from neck, an obtuse canthus. Body cylindrical or slightly depressed, moderately slender; tail short.

Eyes moderate in size; pupils round.

Head scales: The usual 9 crown scales; rostral broad, sometimes pointed and upturned; frontal long, supraoculars large. Laterally, a loreal present, separating nasal and preoculars; suboculars present, separating eye from labials.

Body scales: Dorsals smooth or weakly keeled, with apical pits, in 15–22 oblique rows at midbody, fewer (11–14) posteriorly. Ventrals rounded, 109–155; subcaudals single or paired, 10–33.

VIPERIDAE: Genus *Cerastes Laurenti,* 1768.

Horned vipers.

Two species are recognized; both are restricted to the desert regions of northern Africa and western Asia. Neither is a large species; the bite is painful but usually not serious.

Definition: Head broad, flattened, very distinct from neck; snout very short and broad, canthus indistinct. Body depressed, tapered, moderately slender to stout; tail short.

Eyes small to moderate in size; pupils vertically elliptical.

Head scales: Head covered with small irregular, tubercularly-keeled scales; a large erect, ribbed horn-like scale often present above the eye; no other enlarged scales on crown. Laterally, nasal separated from rostral by 1–3 rows of small scales; eye separated from supralabials by 3–5 rows of small scales.

Body scales: Dorsals with apical pits, large and heavily keeled on back, smaller laterally, oblique, with serrated keels, in 23–35 rows at midbody. Ventrals with lateral keel, 102–165; subcaudals keeled posteriorly, all paired, 18–42.

African Desert Horned Viper, *Cerastes cerastes* (Linnaeus).

Identification: Many individuals of this species have

FIGURE 51.—African Desert Horned Viper, *Cerastes cerastes.* Photo by Zoological Society of San Diego.

a long spinelike horn above the eye; in some, however, this is short or absent. Body form is typically viperine with wide triangular head, thick body, and short tail tapering abruptly behind vent. Top of head covered with small scales; subcaudals paired; ventrals feebly keeled; 15 or more scales across top of head; more than 130 ventrals.

Ground color yellowish, pale gray, pinkish or pale brown with rows of dark brown, blackish or bluish spots that may fuse into crossbars; below whitish, tip of tail black.

Average length 20 to 25 inches; maximum about 30 inches.

Distribution: The Sahara region and Arabian Peninsula; parts of the Middle East.

Inhabits deserts where there are rock outcroppings and fine sand, often in very arid places; however, oases are not avoided. It usually hides in rodent holes and under stones.

Remarks: Chiefly active at night. Like many desert snakes, it often uses the sidewinding type of locomotion. When angered it rubs inflated loops of its body together to make a rasping hiss as does the saw-scaled viper (*Echis*).

It is not a particularly bad tempered or dangerous snake, although it is inclined to stand its ground if disturbed. It causes some snakebite accidents, but fatalities are rare. Antivenin is produced by the Institut Pasteur, Paris, and the Institut Pasteur d'Algerie, Algiers.

Sahara Sand Viper, *Cerastes vipera* (Linnaeus).

Identification: Very similar in appearance to the desert horned viper except that the horns are absent; 9–13 scales across top of head; fewer than 130 ventrals.

Color much as in the horned viper but tending to be more faded with spots less well defined; tip of tail black in female, light in male.

Average length 13 to 18 inches; maximum about 22 inches; females larger than males.

Distribution: Eastern and central Sahara to Israel in sandy desert.

Remarks: Found only in tracts of fine loose sand into which it buries itself when alarmed; usually spends the day buried in sand at the base of a shrub; active at night. In places where this viper is common, the horned viper is rare or absent and vice versa. Care should be taken to differentiate this snake from *Echis carinatus*, a much more dangerous snake.

It is not a very dangerous snake; the venom is small in amount and not highly toxic. Antivenin is produced by the Institut Pasteur, Paris, and by Behringwerke (Polyvalent).

VIPERIDAE: Genus *Echis* Merrem, 1820.

Saw-scaled vipers.

Two species are recognized. One (*E. coloratus*) is restricted to eastern Egypt, the Arabian Peninsula, and Israel. The other (*E. carinatus*) ranges from Ceylon and southern India across western Asia and north Africa southward into tropical Africa. Although neither attains a length of 3 feet, they posses a highly toxic venom and are responsible for many deaths. When disturbed they characteristically inflate the body and produce a hissing sound by rubbing the saw-edged lateral scales against one another. This same pattern of behavior is shown by the nonpoisonous egg-eating snakes *Dasypeltis*.

Definition: Head broad, very distinct from narrow neck; canthus indistinct. Body cylindrical, moderately slender; tail short.

Eyes moderate in size; pupils vertically elliptical.

Head scales: A narrow supraocular sometimes present; otherwise crown covered with small scales, which may be smooth or keeled. Rostral and nasals distinct. Laterally, eye separated from labials by 1–4 rows of small scales; nasal in contact with rostral or separated from it by a row of small scales.

Body scales: Dorsals keeled, with apical pits, lateral scales smaller, with serrate keels, in 27–37 oblique rows at midbody. Ventrals rounded, 132–205; subcaudals single, 21–52.

Saw-scaled Viper, *Echis carinatus* (Schneider).

Identification: Head short and wide, snout blunt; body moderately stout; scales on top of head small, keeled; scales on side of body strongly oblique, the keels with minute serrations; subcaudals single.

Color pale buff or tan to olive brown, chestnut or reddish; midline row of whitish spots; sides with narrow undulating white line; top of head usually shows light trident or arrowhead mark with 3 prongs directed posteriorly and one anteriorly; belly white to pinkish brown stippled with dark gray.

Average length 15 to 20 inches; maximum about 32 inches; sexes of about equal size.

FIGURE 52.—Saw-scaled Viper, *Echis carinatus.* Typical defensive pose. Photo by New York Zoological Society.

Distribution: Almost the entire Afro-Asian desert belt from Morocco and Ghana to the southern provinces of Russian Asia and drier parts of India and Ceylon.

Very adaptable, found from almost barren rocky or sandy desert to dry scrub forest and from seacoast to elevations of about 6,000 feet. Very abundant over much of its range.

Remarks: Almost wholly nocturnal in dry hot weather; occasionally diurnal in cool weather; during rainy season often climbs into bushes. Usually tries to escape when encountered, but is very alert and irritable. Assumes characteristic figure-8 coil, rubbing inflated loops of body together to make a distinctive sizzling noise. Strikes quickly and repeatedly with considerable reach for a small snake.

This little viper is an important cause of snakebite accidents and fatalities almost everywhere that it is found. The venom seems to be unusually toxic for man, and death has been recorded following the bite of a snake 10½ inches long. Hemorrhages, internal and external, are a prominent part of the clinical picture. Serious late complications are frequent, and death may occur 12 to 16 days after the bite.

Saw-scaled viper antivenins are produced by the Institut Pasteur, Paris; Behringwerke, Marburg-Lahn, Germany; Central Research Institute, Kasauli, India; Haffkine Institute, Bombay, India; Tashkent Institute, Moscow; State Razi Institute, Tehran, Iran; and the South African Institute for Medical Research, Johannesburg.

VIPERIDAE: Genus *Vipera* Laurenti, 1768.

True adders.

Eleven species are recognized. This is an especially variable group, with some members that are small and relatively innocuous (e.g., *V. berus*) and others that are extremely dangerous (*V. lebetina*, *V. russelii*). They are found from northern Eurasia throughout that continent and into north Africa. One species ranges into the East Indies (*V. russelii*), and two are found in east Africa (see Remarks under *V. superciliaris*). Both the sunbnosed viper. *V. latasti*, and *V. mauritanica* are found in this region (see p. 74).

Definition: Head broad, distinct from narrow neck; canthus distinct. Body cylindrical, varying from moderately slender to stout; tail short.

Eyes moderate in size to small; pupils vertically elliptical.

Head scales: Variable; one species (*V. ursinii*) has all 9 crown scutes, most species have at least the supraoculars, but even these are absent in one (*V. lebetina*); head otherwise covered with small scales. Laterally, nasal in contact with rostral or separated by a single enlarged scale (the nasorostral), eye separated from supralabials by 1–4 rows of small scales.

Body scales: Dorsals keeled, with apical pits, in 19–31 nonoblique rows at midbody. Ventrals rounded, 120–180; subcaudals paired, 20–64.

Sahara Rock Viper, *Vipera mauritanica* (Gray).

Identification: Closely related to *V. lebetina* of the Near and Middle East. Absence of serrated keels on the lateral scales or keeled ventrals distinguishes it from *Cerastes;* paired subcaudals and lack of serrated keels distinguishes it from *Echis;* a blunt rather than upturned snout distinguishes it from *Vipera latasti;* the lateral position of the nostrils, more slender body and fewer than 27 scale rows at midbody distinguishes it from the puff adder (*Bitis arietans*).

Ground color grayish, reddish, or brown with series of oval or rectangular dark blotches that tend to fuse forming the zigzag stripe of many European and Asian vipers; belly pale extensively clouded with dark gray. Its pattern is much like that of the Palestine (*V. x. palaestinae*) viper (see page 112).

Average length 35 to 45 inches.

Distribution: The northwestern part of the Sahara region from Spanish Sahara to Tripolitania (northwest Libya). Found on hillsides with scrubby vegetation and large flat stones.

Remarks: Hides by day in rock crevices and mine tunnels; most active about twilight.

It is considered a dangerous species. Specific antivenin is produced by the Institut Pasteur d'Algerie, Algiers.

REFERENCES

BONS, J. and B. GIROT. 1962. Cle Illustree des Reptiles du Maroc. Trav. Inst. Sci. Cherifien Ser. Zool. No. 26, p. 1–62, figs. 1–15.

KRAMER, Eugen and H. SCHNURRENBERGER. 1963. Systematik, Verbreitung und Okologie der Libyschen Schlangen. Rev. Suisse de Zool., vol. 70, pp. 453–568, pls. 1–4, figs. 1–13.

MARX, Hymen. 1956. Keys to the Lizards and Snakes of Egypt. Research Rpt. NM 005 050.39.45, NAMRU-3, Cairo pp. 1–8.

SAINT-GIRONS, H. 1956. Les Serpents du Maroc. Var. Scient. Soc. Sci. Nat. Psyc. Maroc, vol. 8, pp. 1–29, pls. 1–3, figs. 1–9.

VILLIERS, Andre. 1950. Contribution a l'etude du peuplement de la Mauritanie. Ophidiens. Bull. Inst. Francais d'Afrique Noire, vol. 12, pp. 984–998, figs. 1–2, tables.

Section 6

CENTRAL AND SOUTHERN AFRICA

Definition of the Region:

All of Africa south of the Sahara Desert region. The northern border of this huge area coincides with the southern boundaries of Mauritania, Mali, Niger and Chad; and with the northern boundary of the Sudan. Madagascar, off the east coast, has no venomous snakes.

TABLE OF CONTENTS

TABLE 10.—DISTRIBUTION OF POISONOUS SNAKES OF CENTRAL & SOUTHERN AFRICA

	Republic of South Africa (Inc. Basutoland; Swaziland)	Rhodesia	Botswana (Bechuanaland)	S. W. Africa	Angola	Zambia	Mozambique—Malawi	Nyasaland	Zanzibar Is.	Tanzania	Kenya	Uganda	Rwanda & Burundi	Congo, Rep. of (former Belgian Congo)	Congo Republic (former French Congo)	Gabon—Rio Muni	Central Africa Republic	Cameroon	Nigeria	Dahomey	Togo	Upper Volta	Ghana (G. Coast)	Ivory Coast	Liberia	Sierra Leone	Guinea—Port. Guinea	Senegal—Gambia	Somalia	Ethiopia—Fr. Somaliland	Sudan
COLUBRIDAE																															
Dispholidus typus	X	X	X	X	X	X	X	X	X	X	X	X	X	X			?						X	X				?	W		
Thelotornis kirtlandii	X	X		N	X	X	X	X		X	X	?		X			?						X	X					S		
ELAPIDAE																															
Aspidelaps lubricus	X		X	X	S																										
A. scutatus	X	X	X	X			X																								
Boulengerina annulata								?		X				X	?	X	?	X	?												
B. christyi														W																	
Dendroaspis angusticeps	E	E				X	S	?	X	X	X		X	E	?		?	X	?	?		?	?	?	?	X	?				
D. jamesoni					X						W	?	X	E		?	?	N	N	X	X	?	X	X	?		X	?	N	X	X
D. polylepis	X	X	X	X	X	X	S	X	X	X	X	?	X				?	N					X	X				X	S	X	X
D. viridis												X	X		?	?		X	X	X	X	?	X	X	X	X	X	X			
Elaps dorsalis																	?														
E. lacteus	X																?									X					
Elapsoidea sundevallii	X		X	X	X	X	S	X	X	X	X	X	X	X		?	?	N	N	X	X	?	X	X		X	X	X	SW	S	X
Hemachatus haemachatus	X	X	X				N										?														
Naja haje	X	X	X	X	X	X	S	X	X	X	X	X	X	NE			?	N	N	X	X	?	X	X		X	X	X	N	S	X
N. melanoleuca	E	E	X	X	X	X	S	X	X	X	X	X	X	X		?	?	X					X	X	X	X	X	X	S	X	
N. nigricollis	X	X	X	X	X	X	X	X	X	X	X	X	X	X	?		?	X	X	X	X		X	X		X	X	X	X		X
N. nivea																	?														
Paranaja multifasciata														X		X	?	X													
Pseudohaje goldii	X	X	X	X	X		S					X		X	?	?	?	?	X	X	X	?	X	?	?	X	X		SW		
P. nigra					?	?	N						?	X	?		?						?	?							
VIPERIDAE																															
Adenorhinos barbouri										X																					
Atheris ceratophorus										X																					
A. chloroechis															?	X		X	X		X		X	X	X	X	X				
A. hispidus												X	X	E		?															
A. katangensis												SW	X	X																	
A. nitschei					?	X	S	X		W	?	X	X	E	?	?	?	?					?							?	
A. squamigera				X	X					W	?	X	?	X	X	X	?	X												?	

(Continued on page 88)

INTRODUCTION

The poisonous snake fauna of central and southern Africa is a large and diverse one. There are no venomous terrestrial snakes on Madagascar, off the east coast, and only occasionally does a lone sea snake (*Pelamis*) wash ashore there or along the eastern coast of the mainland. However, there are records of sea snakes from as far south and west as Capetown, although there appear to be no reports of any person having been bitten in African waters by sea snakes.

Other than the sea snakes, the African poisonous snakes belong to three families, the Colubridae, the Elapidae, and the Viperidae. Africa is the only region where colubrid snakes are considered dangerously venomous, but here there are two tree snakes, the boomslang (*Dispholidus*) and the bird snake (*Thelotornis*), that have proven to be capable of inflicting lethal bites.

The elapids include burrowing snakes, some of which (e.g., *Elaps*) are so small as to be of little concern. However, there are many dangerous terrestrial species as well as a number of specialized arboreal kinds (*Pseudohaje, Dendroaspis*). The most terrestrial of the mambas, *Dendroaspis polylepis*, the black mamba, attains a length of about 14 feet and is one of the most dangerous snakes in existence. Other especially dangerous terrestrial species are the Egyptian cobra (*Naja haje*), which has a wide range through central Africa, the spitting cobra (*Naja nigricollis*), also with a wide range, and the yellow cobra (*N. nivea*) and ringhals (*Hemachatus*) of southern Africa.

The vipers are an equally diverse group. A genus of burrowing mole vipers (*Atractaspis*) is found throughout the region. Even though most of these do not exceed 2 feet in length, they are capable of inflicting dangerous bites. Some of the central African terrestrial vipers are the largest members of their family: the massive Gaboon viper (*Bitis gabonica*) exceptionally attains a length of 6 feet, with fangs almost 2 inches long. In addition there are relatively small desert vipers in the temperate south. However, the most widespread, the most commonly seen, and probably the greatest killer of man is the common puff adder (*Bitis arietans*). The bush vipers (*Atheris*) do not appear to be an important danger.

With such a wealth of dangerously venomous snakes, one would expect snakebite to be an important cause of death in Africa. However, the few statistics available do not give this impression. The reported incidence of death from

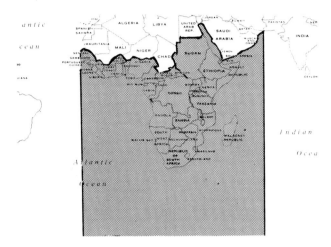

Map 7.—Section 6, Central and Southern Africa.

snakebite is much lower than in the tropical countries of the Asian mainland. Whether this is a true picture or if it is distorted by poor reporting is as yet unknown.

The vipers of the genera *Bitis, Echis, Atheris*, and *Vipera* have the common attributes of venomous snakes—broad distinct head and eyes with vertically elliptical pupils. However, this is not true of the night adders (*Causus*), the mole vipers (*Atractaspis*), or the various elapid and dangerous colubrid species. These have no general characteristics that set them off from harmless snakes. However, poisonous snakes make up less than a quarter of the snake fauna throughout the region and it is not too difficult to learn the venomous kinds in any one area.

Many of the elapid species are cobras (*Naja*) or cobra-like kinds, and while a cobra minding its own business looks very much like any other snake, a disturbed cobra will quickly spread a hood—which is a plain and distinctive warning. Even some of the elapids without well-developed hoods (e.g., the mambas, *Dendroaspis*) will flatten the neck if disturbed, and some which do not resemble cobras in any way (e.g., *Aspidelaps*) will flatten the neck and raise the anterior part of the body in the familiar cobra stance.

TABLE 10.—DISTRIBUTION OF POISONOUS SNAKES OF CENTRAL & SOUTHERN AFRICA (continued)

VIPERIDAE (continued)	Sudan	Ethiopia–Fr. Somaliland	Somalia	Senegal–Gambia	Guinea–Port. Guinea	Sierra Leone	Liberia	Ivory Coast	Ghana (G. Coast)	Upper Volta	Togo	Dahomey	Nigeria	Cameroon	Central Africa Republic	Gabon–Rio Muni	Congo Republic (former French Congo)	Congo, Rep. of (former Belgian Congo)	Rwanda & Burundi	Uganda	Kenya	Tanzania	Zanzibar Is.	Nyasaland	Mozambique–Malawi	Zambia	Angola	S.W. Africa	Botswana (Bechuanaland)	Rhodesia	Republic of South Africa (Inc. Basutoland; Swaziland)
Atractaspis aterrima					X			X	X						?	X	?	X	?	X		X									
A. battersbyi		X	S															X	?	?											
A. bibronii		X														X	?	X		?	X	X	X	X	S	X	X	X	X	X	X
A. boulengeri																		X													
A. coalescens											X	X														X	X				
A. congica														X				X													
A. corpulenta							X	X	X		?	?	?	X	?	X	?	X													
A. dahomeyensis					N	?	?	X	X		?	?	?	X	?																
A. duerdeni														X															X		
A. engdahli					N		X							X													N				
A. irregularis	X				X	?		?	?	?	?	?	X	X	?	?	?	E	X	X	X	NW									
A. leucomelas		X	X																												
A. microlepidota	X	X	X	X				?					N	?		?		X			X										
A. reticulata														?		X		X													
A. scorteccii			N																												
Bitis arietans	X	X	X	X	X	?	X	X	X	?	?	?	?	X	?	?	?	E	X	X	X	X	?	X	X	X	X	X	X	X	X
B. atropos																														E	X
B. caudalis																												X	X	W	SW
B. cornuta																					?							S	?		E
B. gabonica	S				X	?	X	X	X					X	?	X	?	X	X	X	?	E	?	X	N	X	X			E	
B. heraldica																											X				
B. inornata																															SE
B. nasicornis	S				X	?	X	?	X					X	?	X	?	X	?	X							X				
B. peringueyi																											X	S			
B. worthingtoni																					W										
Causus defilippii																				?					N						
C. lichtensteinii														X				X		X									S		
C. resimus																		X		X	?										
C. rhombeatus	X	?	S		X	?	X										?	E		X	W	X	X	X	X	X	X		X	X	X
Echis carinatus	X	?	W	X	X			X	X					X	?	X		X	X	X	N	X		X	S	X	X	X	X	X	X
Vipera hindii																					C	X				X	X				
V. superciliaris																					C				C						

The symbol X indicates distribution of the species is widespread within the area. Restriction of a species to part of a unit is indicated appropriately (SW = Southwest, etc.). The symbol ? indicates suspected occurrence of a species within the unit without valid literature.

KEY TO GENERA

1. A. Crown of head covered with small irregular scales;
      pupil of eye vertically elliptical_____ 15
   B. Crown of head with nine large regular plates; pupil of
      eye various shapes_____ 2
2. A. Loreal scale absent; preoculars in contact with nasal or
      separated by downward extension of prefrontal_____ 6
   B. Loreal scale(s) present, separating preoculars from
      nasal_____ 3
3. A. Pupil of eye horizontally elliptical_____ *Thelotornis*
   B. Pupil of eye round_____ 4
4. A. Eye separated from supralabials by row of subocular
      scales_____ *Causus*
   B. Eye in contact with supralabials_____ 5
5. A. Dorsals distinctly keeled, in 17–21 rows at midbody_____ *Dispholidus*
   B. Dorsals smooth, in 13–15 rows at midbody_____ *Pseudohaje*
6. A. Preoculars (3) widely separated from nasal; prefrontals
      expanded laterally to touch labials_____ *Dendroaspis*
   B. Preoculars (1 or 2) in contact with nasal_____ 7
7. A. Rostral very large, concave below, separated from other
      scales on sides_____ *Aspidelaps*
   B. Rostral not concave below, not separated from other
      scales_____ 8
8. A. Dorsals distinctly keeled_____ *Hemachatus*
   B. Dorsals perfectly smooth_____ 9
9. A. Eye very small; frontal more than twice as broad as
      supraoculars_____ *Atractaspis*
   B. Eye small to very large; frontal not twice as broad as
      supraoculars_____ 10
10. A. Tail moderate to long; more than 41 subcaudals_____ 13
    B. Tail short; fewer than 42 subcaudals_____ 11
11. A. Rostral enlarged, obtusely pointed; dorsals in 13 rows
       at midbody_____ *Elapsoidea*
    B. Rostral normal, rounded; dorsals 15–17 rows at midbody_____ 12
12. A. Anal plate divided; dorsals 15 throughout_____ *Elaps*
    B. Anal plate entire; dorsals 15–17 at midbody, more on
       neck, fewer posteriorly_____ *Paranaja*
13. A. Eye very large; dorsals 13–15 rows at midbody _____ *Pseudohaje*
    B. Eye small to moderate; dorsals 17 or more at midbody_____ 14
14. A. Dorsal pattern of 3–24 distinct dark crossbars on lighter
       ground color; 3–4 small teeth on maxillary bone____ *Boulengerina*
    B. No such pattern; 0–2 (rarely 3) small teeth on maxillary
       bone_____ *Naja*
15. A. Lateral scales with serrate keels_____ 18
    B. Lateral scales not serrately keeled_____ 16
16. A. Subcaudals paired_____ 19
    B. Subcaudals single_____ 17
17. A. Fewer than 30 subcaudals; fewer than 130 ventrals_____ *Adenorhinos*
    B. More than 30 subcaudals; more than 130 ventrals_____ *Atheris*

KEY TO GENERA (Continued)

18. A. Ventrals with lateral keel; subcaudals paired_____ *Cerastes*
    B. Ventrals rounded; subcaudals single_____ *Echis*
19. A. Rostral in contact with nasal, or separated from it by a
    single large scale_____ *Vipera*
    B. Rostral separated from nasal by 1 or more rows of
    small scales_____ *Bitis**

*B. worthingtoni, with the nasal in contact with rostral, single subcaudals, and lateral keels on the ventrals, will not key out properly.

GENERIC AND SPECIES DESCRIPTIONS

COLUBRIDAE: Genus *Dispholidus* Duvernoy, 1832.

Boomslang.

A single species, *D. typus* Smith. This snake, found only in tropical and southern Africa, is the most dangerous member of the family Colubridae.

Definition: Head oval but distinct from slender neck; crown of head convex. Snout short with a distinct canthus. Body slender and elongate, moderately compressed; tail long and slender.

Eyes very large; pupils round.

Head scales: The usual 9 on the crown. Laterally, a single loreal scale separates the nasal from the one or two preoculars.

Body scales: Dorsals narrow, distinctly keeled, and with apical pits; in 17–21 oblique rows at midbody, more (21–25) anteriorly, fewer (13) posteriorly. Ventrals of normal size, obtusely angulate laterally, 164–201; anal plate divided (Like most "present or absent" scale characteristics, this is not true 100 percent of the time; the anal plate is rarely entire. The question of identification of a boomslang with an entire anal plate caused the death of a noted herpetologist, Karl P. Schmidt, in 1957. *See* Pope, 1958); subcaudals paired, 87–131.

Maxillary teeth: A series of 7–8 small subequal teeth followed, after a short interspace, by 3 very long grooved fangs.

Boomslang, *Dispholidus typus* Smith.

Identification: The boomslang does not look much different from many other tree-dwelling snakes which inhabit its range and, of course, it is not always in a tree. However, the innocuous green bush snakes (*Philothamnus*) have smooth scales, and keeled and notched ventrals, while the dangerous mambas (*Dendroaspis*) have a much longer and narrower head, lack a loreal scale, and have smaller eyes. Adult boomslangs average 4 to 5 feet, with the record length being "a little over 6 feet."

Color varies from almost black to almost unicolor green; no blotches or distinct spots. Individual dorsal scales may be yellow, brown, or green, often with black on the margins. Ventrals black to greenish white, depending on dorsal color. No distinct head pattern.

Distribution: Open savannah and brushy country throughout tropical and southern Africa; not found in

FIGURE 53. Boomslang, *Dispholidus typus*. Photo by Roy Pinney and National Zoo, Washington, D.C. (See also plate VII, figures 1, 4; plate VIII, figure 2.)

rain forest regions nor in true desert.

Remarks: This snake is not aggressive and will quickly make for the nearest tree or bush if surprised on the ground. In its arboreal habitat it disappears quickly. However, if cornered or restrained, it inflates its neck to more than double its normal dimensions. This exposes the skin between the scales of that region, which is often brightly colored. If its bluff is unsuccessful, the boomslang will bite.

Although it is a rearfanged colubrid snake, the boomslang has relatively long fangs and its venom, though in small quantities, is more toxic, drop-for-drop, than that of African cobras and vipers. The venom causes severe internal bleeding; every mucous membrane may ooze blood; a number of deaths have been reported. (Pope, 1958.)

A specific antivenin (Boomslang) is produced by the South African Institute for Medical Research, but it is in short supply and is usually held by them for severe cases that come to their attention.

COLUBRIDAE: Genus *Thelotornis* Smith, 1849.

Bird snake.

A single species, *T. kirtlandii* (Hallowell). The bird snake is restricted to tropical and southern Africa. Other than the boomslang of the same region, it is the only species of colubrid, rear-fanged snake that is known to cause serious injury, and occasionally death.

Definition: Head elongate, flattened and distinct from neck; a distinct and projecting canthus which forms a shallow groove below it on the side of the snout. Body slender and elongated, cylindrical; tail long.

Eyes large; pupils horizontally elliptical (keyhole-shaped).

Head scales: The usual 9 on the crown; internasals large; parietals bordered posteriorly by 3 large scales. Laterally, 1–3 loreal scales separate the nasal from the preocular.

Body scales: Dorsals narrow, feebly keeled, with apical pits, in 19 oblique rows anteriorly and at midbody, fewer (11–13) posteriorly. Ventrals rounded, 147–189; anal plate divided; subcaudals paired, 131–175.

Maxillary teeth: A series of 11–16 small teeth which gradually increase in length followed, after a short interspace, by 3 long grooved fangs.

Bird Snake, *Thelotornis kirtlandii* (Hallowell).

Identification: This slender-snouted tree snake is most easily recognized by its long, flat-crowned head with shallow lateral grooves that extend forward from the eyes. Its eyes are large and have horizontally elliptical pupils. There are usually two loreals, one behind the other, and the scales on the sides of the body

FIGURE 54.—Bird Snake, *Thelotornis kirtlandii*. Photo by Zoological Society of San Diego. (See also plate VIII, fig. 1.)

are long and narrow, almost rectangular in shape. Adults average about 4 feet; record length 5½ feet.

Body ashy gray to pinkish brown above, unicolor or with poorly-distinguished blotches and crossbands. Underneath, the color is brownish or grayish, heavily speckled with brown. Head unicolor green, pinkish, or purplish brown above, flecked with dark brown or black; occasionally a Y-shaped design on back of head; a dark band extending from behind eye obliquely onto neck.

Distribution: The tropical forests and savannah regions of central and southern Africa, southward to the Transvaal in the east and to central South-West Africa in the west.

Remarks: This snake seldom attempts to bite; nevertheless, its highly toxic venom has caused a few fatalities. When molested it assumes a threatening attitude and inflates its neck greatly, mainly in a vertical direction. This brings to view a bold pattern of black crossbands on a light background.

No antivenin is produced for this snake.

ELAPIDAE: Genus *Aspidelaps* Fitzinger, 1843.

Shield-nose snakes.

Two species are recognized; both are restricted to southern Africa. They are small semiburrowing snakes with a specialized snout. Although they possess relatively large fangs, neither species attains a length of over 30 inches and they are not considered dangerous.

Definition: Head short and only slightly distinct from neck; a broad snout modified for burrowing; canthus indistinct. Body cylindrical or somewhat depressed, stout; tail short, obtusely pointed.

Eyes moderate in size; pupils round or vertically elliptical.

Head scales: The usual 9 on the crown; rostral very large, concave below, curved backward over snout, separated from other scales on sides; prefrontals very short. Laterally, nasal in broad contact with single preocular.

Body scales: Dorsals smooth or faintly keeled (in *A. scutatus*) in 19–23 oblique rows anteriorly and at midbody, fewer (15) posteriorly. Ventrals 115–172; anal plate entire; subcaudals paired, 20–38.

Maxillary teeth: Two rather large tubular fangs with external grooves; no other teeth on the bone.

ELAPIDAE: Genus *Boulengerina* Dollo, 1886.

Water cobras.

Two species are recognized; they are found in central Africa from Nyasaland to the Congo region. They are large snakes, attaining lengths of over 8 feet. They are not aggressive but are considered dangerous.

Definition: Head short, distinct from neck; an indistinct canthus. Body cylindrical and moderately slender; neck capable of being spread into a hood; tail of moderate length, tapering.

Eyes small; pupils round.

Head scales: The usual 9 on the crown; frontal small. Laterally, nasal in contact with single preocular.

Body scales: Dorsals smooth, in 17–23 oblique rows at midbody, the same number or more (17–25) anteriorly, fewer (13–17) posteriorly. Ventrals 192–227; anal plate entire; subcaudals paired, 67–80.

Maxillary teeth: Two large tubular fangs with external grooves followed, after an interspace, by 3–4 small teeth.

Banded Water Cobra, *Boulengerina annulata* (Buchholz and Peters).

Identification: This is a large fish-eating cobra that is always found in or near water. It is especially common along some of the shores of Lake Tanganyika. The western race (*B. a. annulata*) has a series of 21–24 narrow black crossbands on a brown or tan background; the eastern subspecies (*B. annulata stormsi* Dollo) has only 3–5 such bands on the neck; the remainder of the

FIGURE 55.—Banded Water Cobra, *Boulengerina annulata*. This subspecies, *B. a. stormsi*, has only a few black bands on the neck; the western form, *B. a. annulata*, has bands throughout the body. Photo by Zoological Society of San Diego. (See also plate VII, figure 9.)

body is unicolor brown. The nonvenomous watersnake *Grayia*, which has the same range, looks much like the western form of the water cobra but may be distinguished by the presence of a loreal scale. Although it may spread the body, *Grayia* does not have a hood. Water cobras raise the anterior part of the body and spread a narrow hood as a threat, in typical cobra fashion. Adults average 5 to 7 feet in length; record length is about 9 feet.

Dorsal scales 21–23 at midbody, more (23–25) on the neck, fewer (15–17) posteriorly. Ventrals 192–227; subcaudals 67–80.

Distribution: Nyasaland and Lake Tanganyika westward through the rain forest regions to the western Congo and Cameroon.

Remarks: These large water cobras are not aggressive and appear to offer little danger to persons that leave them alone. Little is known of the effect of their bite and no antivenin is produced for the snakes of this genus.

ELAPIDAE: Genus *Dendroaspis* Schlegel, 1848.

Mambas.

Four species are currently recognized. They range over most of central and southern Africa. Due to their size, speed, and highly toxic venom, they are considered among the most dangerous of all snakes. The fact that all are greenish when young has confused the identity of these snakes for many years. One species, *D. polylepis*, attains a length of 14 feet.

Definition: Head narrow and elongate, slightly distinct from neck; a distinct canthus. Body slender and tapering, slightly compressed; neck may be flattened when snake is aroused, but there is no real hood; tail long and tapering.

Eyes moderate in size; pupils round.

Head scales: The usual 9 on the crown; frontal broad anteriorly, narrow posteriorly. Laterally, nasal widely separated from preoculars by prefrontal.

Body scales: Dorsals smooth and narrow, in 13–25 distinctly oblique rows at midbody, the same or more rows anteriorly, fewer posteriorly. Ventrals 201–282; anal plate divided; subcaudals paired, 99–131.

Maxillary teeth: Two large tubular fangs without external grooves; no other teeth on bone.

Eastern Green Mamba, *Dendroaspis angusticeps* (Smith).

Identification: This is a long and very slender bright green tree snake that is often confused with the rear-fanged boomslang, *Dispholidus*, and the harmless green bush-snakes, *Philothamnus*. It can be distinguished from both by the smaller eyes and by the absence of the loreal scale, and from the bush-snakes by the absence of keels and notches on the ventral plates. It differs from the black mamba (*D. polylepis*), the only other mamba in its range, by its bright green color, the light color (white to bluish white) of the inside of its

FIGURE 56.—Eastern Green Mamba, *Dendroaspis angusticeps*. The bright green color and the long head distinguish this species. Photo by New York Zoological Society.

mouth, and the fewer dorsals and ventrals. Adults average 6–8 feet in length; record length is about 9 feet.

Dorsals in 17–19 rows at midbody; ventrals 201–232; subcaudals 99–126.

Distribution: A narrow range in the forests and brushy country of east Africa from Kenya southward to southern Natal and northeastern Cape Province. It is found on the island of Zanzibar.

Remarks: This species and the more dangerous black mamba were confused for many years. The green mamba is much more arboreal, seldom found on the ground. It is shy and avoids man if possible. Its venom differs in many ways and is only about half as toxic as that of the black mamba.

A polyvalent antivenin (mamba) is produced by the South African Institute for Medical Research, Johannesburg.

Jameson's Mamba, *Dendroaspis jamesoni* (Traill).

Identification: A mainly green tree snake with scales usually edged in black, the overall coloration becoming darker posteriorly, with the tail entirely black in some individuals. It differs from the harmless bush-snakes (*Philothamnus*), and from the rear-fanged boomslang (*Dispholidus*), with which it may be confused, in lacking the loreal scale and in having smaller eyes; and from the bush snakes too in the darker coloration and the absence of lateral keels and notches on the ventrals. It differs from the black mamba (*D. polylepis*) in having black edging on the scales and fewer dorsals and ventrals. Adults average 6 to 7 feet in length; a record is 8 feet, 1 inch (Schmidt, 1923: 131).

Dorsals in 15–19 rows at midbody, the same number or more (15–19) on the neck, fewer, (11–13) posteriorly. Ventrals 210–239; subcaudals 99–121.

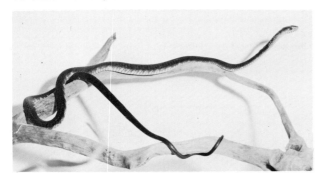

FIGURE 57.—Jameson's Mamba, *Dendroaspis jamesoni.* This individual demonstrates the typical alert pose of mambas. Photo by New York Zoological Society. (See also plate VII, figure 10.)

Distribution: The tropical rain forest region from western Kenya and Tanganyika to Guinea and Angola.

Remarks: This mamba is found both on the ground and in bushes; however little has been reported on its habits.

A polyvalent antivenin for its bite (mamba) is prepared by the South African Institute for Medical Research, Johannesburg.

Black Mamba, *Dendroaspis polylepis* Günther.

Identification: Adult snakes are olive brown to dark gunmetal gray. However, hatchlings are grayish green or olive and this has caused the black mamba to be con-fused with the eastern green and Jameson's mambas that share parts of its range. The canthus is particularly sharp in the black mamba and the head is impressively high in large individuals. This large, relatively slender, and very fast-moving snake is not readily confused with any nonvenomous species. It differs from other mambas in being darker, in having a bluish gray to blackish color inside the mouth, and more dorsal scale rows and more ventral scutes. The forest cobra, *Naja melanoleuca*, differs in having a prominant hood and very glossy scales. Adults average 9 to 10 feet; the record length is about 14 feet.

Dorsals in 21–25 rows at midbody, the same number or more on the neck (25), fewer (15–19) posteriorly. Ventrals 242–282; subcaudals 105–131.

FIGURE 58.—Black Mamba, *Dendroaspis polylepis.* The sharply-defined canthus rostralis is plain here. Photo by New York Zoological Society. (See also plate III, figure 1; plate VII, figure 11.)

Distribution: Inhabits low-lying (below 4,000 feet) open bush country from Ethiopia and Somalia, avoiding the western rain forest region, southward to Natal and South-West Africa.

Remarks: This snake is found in trees and bushes less often than the other mambas. It is one of the fastest snakes known, and has been clocked at slightly over 7 miles per hour, or perhaps twice as fast as the fastest North American snake. It gives the impression of great speed and in some of the older literature it was reported to "exceed the speed of a running horse." A recent publication estimates the speed at "probably not exceeding 20 mph."

It is certainly one of the most dangerous snakes now living. Although it ordinarily makes for its hole when disturbed, it is ready to fight if suddenly disturbed. The typical attitude of alert defense is with the head raised well off the ground, mouth slightly agape (showing the black lining) and tongue flicking rapidly from side to side. No other mamba shows such peculiarities. When angered, the snake emits a hollow-sounding hiss and spreads its neck. It is said to strike out for 40 percent of its length; the average snake strikes out for 25 to 30 percent.

A large black mamba secretes enough venom to kill 5 to 10 men and few people survive its bite unless antivenin is administered promptly. The venom inhibits breathing and apparently also inhibits the branch of the vagus nerve that controls heartbeat, this causes the heart to beat wildly.

A polyvalent antivenin (mamba) is produced by the South African Institute for Medical Research, Johannesburg.

Western Green Mamba, *Dendroaspis viridis* (Hallowell).

Identification: This is another of the arboreal mambas. Like many forest snakes it has an overall green or yellowish color, but each of the dorsal scales, as well as the head scales, is edged with black. The dorsals are extremely large and narrow; each dorsal except the one bordering the ventral row is equal to two ventrals in length. This snake has fewer dorsal rows

FIGURE 59.—Western Green Mamba, *Dendroaspis viridis.* The large oblique black-bordered scales distinguish this species. Photo by New York Zoological Society.

than any of the other snakes with which it might be confused and also lacks the loreal scale typical of colubrid snakes. No other mamba occurs within its range. Adults average 6 to 7 feet in length.

Dorsals in 13 rows at midbody, more (15) on the neck, fewer (9) posteriorly. Ventrals 211–225; subcaudals 105–119.

Distribution: The tropical rain forest areas of the western bulge of Africa; from the Senegal to the Niger, also the island of Sao Tome.

Remarks: Little appears to be known of the habits of this west African mamba.

A monovalent antivenin ("Dendraspis") is produced by the Institut Pasteur, Paris.

ELAPIDAE: Genus *Elaps* Schneider, 1801.

African dwarf garter snakes.

Two species are recognized; both are confined to South Africa. One of the species (*E. lacteus*) attains a length of about 2 feet but neither it nor its smaller relative is considered dangerous.

Definition: Head small, not distinct from neck; no canthus. Body slender and cylindrical; tail short.

Eyes small; pupils round.

Head scales: The usual 9 on the crown; frontal long and narrow, internasals short; rostral broad and rounded. Laterally, nasal in narrow contact with single preocular.

Body scales: Dorsals smooth, in 15 rows at midbody. Ventrals 160–239; anal plate divided; subcaudals paired, 25–41.

Maxillary teeth: Two proportionately large tubular fangs without external grooves; no other teeth on the bone.

ELAPIDAE: Genus *Elapsoidea* Bocage, 1866.

African garter snake.

A single species (*E. sundevallii*) with 11 geographic races is currently recognized. It ranges over most of tropical and southern Africa except for the Cape region. It attains a length of 3 to 4 feet and is potentially dangerous. However, it is sluggish and inoffensive and bites only in self-defense.

Definition: Head of moderate size, not distinct from neck; an indistinct canthus. Body moderately slender, cylindrical; tail very short.

Eyes small; pupils round.

Head scales: The usual 9 on the crown; rostral enlarged, obtusely pointed; internasals short. Laterally, nasal in narrow contact with single preocular.

Body scales: Dorsals smooth and rounded, in 13 rows at midbody. Ventrals 138–184; anal plate entire; subcaudals paired (a few sometimes single), 13–19.

Maxillary teeth: Two large tubular fangs with external groove followed, after an interspace, by 2–4 small teeth.

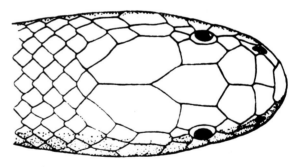

FIGURE 60.—Head Scales of African Garter Snake, *Elapsoidea sundevallii.* Note the broad rostral and short internasals. (See also plate VIII, fig. 3) Redrawn from Pitman, 1938.

ELAPIDAE: Genus *Hemachatus* Fleming, 1822.

Ringhals.

A single species is recognized; it is confined to southern Africa. It is a highly developed "spitting" cobra and is a dangerous species.

Definition: Head rather broad, flattened, not distinct from neck; distinct canthus; snout obtusely pointed. Body moderately slender, slightly depressed, tapering; neck region capable of being expanded into hood; tail moderately long.

Eyes moderate in size; pupils round.

Head scales: The usual 9 on the crown; rostral large and obtusely pointed. Laterally, nasal in contact with single preocular.

Body scales: Dorsals distinctly keeled, in 19 oblique rows at midbody, fewer (15) posteriorly. Ventrals 116–150; anal plate entire; subcaudals 33–47, the first 3–4 frequently single, the remainder paired.

Maxillary teeth: Two short tubular fangs with external grooves; no other teeth on bone.

Ringhals, *Hemachatus haemachatus* (Lacépède).

Identification: A cobra with strongly keeled scales. When the snake raises the anterior part of its body and spreads the hood, as it does in a defensive attitude, it

FIGURE 61.—Ringhals, *Hemachatus haemachatus*. The strongly keeled scales distinguish this species from other cobras. Photo by Zoological Society of San Diego.

exposes a black throat with 1–3, usually 2, light bands on the ventral surface below the hood. The first light band is narrow (1–2 ventrals in width) while the other is broad (5–7 ventrals). Adults average 3½ to 4 feet in length; record length "just over 5 feet" (FitzSimons, 1962:288).

The dorsal color is usually dark brown with irregular crossbands of lighter brown, often with small black spots; occasionally gray or greenish; old individuals become almost unicolor black.

Distribution: Veldt and open country in southeastern and southern Africa from Rhodesia to the southern Cape Province.

Remarks: This is the most highly specialized of the "spitting" cobras. Its fangs are relatively short but the small venom orifice on the front of the fang and

strong muscles around the venom gland allow the ejection of venom in a fine spray to a distance of 5 to 7 feet. The venom is ordinarily aimed at the eyes of the enemy. It causes intense pain and spasm of the eyelids. Destruction of eye tissue and blindness may result if the eye is not washed out immediately with some harmless fluid (Fitzsimons, 1962: 290). The ringhals bites if restrained, and can cause death.

Polyvalent antivenin "Polyvalent" and "Tropical" are produced by the South African Institute for Medical Research, Johannesburg.

ELAPIDAE: Genus *Naja* Laurenti, 1768.

Cobras.

Six species are recognized; all are African except the Asiatic Cobra, *Naja naja*, and range throughout the African continent except for the drifting sand areas of the Sahara region. They are snakes of moderate (4 feet) to large (8 feet) size, with large fangs and toxic venom. The species, *N. nigricollis* "spits" its venom at the eyes of an aggressor; it is found in the southern part of the region of north Africa. The Egyptian cobra (*Naja haje*) and the western subspecies of the Asiatic cobra (*Naja naja oxiana*) are found in the Near and Middle East region.

Definition: Head rather broad, flattened, only slightly distinct from neck; snout rounded, a distinct canthus. Body moderately slender, slightly depressed, tapered; neck capable of expansion into hood; tail of moderate length.

Eyes moderate in size; pupils round.

Head scales: The usual 9 on the crown; frontal short; rostral rounded. Laterally, nasal in contact with the one or two preoculars.

FIGURE 62.—Egyptian Cobra, *Naja haje*. This dangerous species is widespread through the region. Photo by New York Zoological Society. (See p. 80, fig. 49 and plate VIII, figure 7.)

Body scales: Dorsals smooth, in 17–25 oblique rows at midbody, usually more on the neck; fewer posteriorly. Ventrals 159–232; anal plate entire; subcaudals 42–88, mostly paired.

Maxillary teeth: Two rather large tubular fangs with external grooves followed, after an interspace, by 0–3 small teeth.

Forest Cobra, *Naja melanoleuca* Hallowell.

Identification: This large dark terrestrial cobra is most easily recognized by the highly polished dorsal scales and the creamy-white labial scutes which are edged with black. It is sometimes mistaken for the black mamba (*Dendroaspis polylepis*) which does not enter the rain forest region except along its edges. However, the forest cobra is a slower-moving, thicker-bodied, and broader-hooded snake than the mamba. Adults average 6 to 7 feet in length; record length 8 feet 7½ inches (Pitman, 1938 : 215).

This cobra is often unicolor glossy black above. However, the head and sometimes the anterior part of the body may be brown; in young individuals small white spots are scattered or appear as narrow crossbands over the posterior part of the body. Chin and belly creamy white, usually with one or two relatively narrow (4 ventrals wide) black bands under the hood; increasing amounts of black posteriorly (See plate VII, fig. 3; plate VIII, fig. 8).

Dorsals in 17–21 rows at midbody, more (23–29) on the neck, fewer (13) posteriorly. Ventrals 197–226; subcaudals 57–74.

Distribution: Tropical rain forest and subtropical forest areas (and where such forests have recently disappeared) through most of west and central Africa; southward to Angola and Zululand.

Remarks: The forest cobra has a long wedge-shaped hood like that of the spitting cobra (*N. nigricollis*) and is often mistaken for the dark color-phase of the latter. However, it does not "spit" and differs from the spitting cobra in labial color and in the width of the neck bands (4 ventrals versus 7 in *N. nigricollis*).

The forest cobra is seldom aggressive and few bites are reported. However, it has a highly toxic venom and fatalities are known.

A polyvalent antivenin ("Kobra") is produced by Behringwerke, Germany, and Institut Pasteur, Paris.

Spitting Cobra, *Naja nigricollis* Reinhardt.

Identification: A broad black band (width of 7 ventrals or more) under the hood or an entirely black underside, together with the absence of distinctive labial coloration, are the best identification features of this cobra. Its scales are smooth but not so glossy as those of the forest cobra. As in the latter, the hood is long and narrow. Adults average 5 to 6 feet; record is 7 feet, 4 inches.

Body color highly variable, ranging from pinkish-tan in some areas to unicolor black in others. In Southwest Africa there is one race with alternating rings of brown and black. Light areas underneath are often pinkish, even in black individuals (see plate VIII, fig. 9).

Dorsals in 17–25 rows at midbody, more (19–29) on neck, fewer (11–16) posteriorly. Ventrals 176–232; subcaudals 56–73.

Distribution: Throughout the savannah areas of Africa south of the Sahara, also invading newly-cleared areas. From west Africa and southern Egypt, avoiding the dense forests, to the borders of the Cape Province.

Remarks: This is one of the common cobras of the open grasslands and one of the most dangerous snakes of Africa. Although it seldom bites, a large individual can "spit" (actually squirt) its venom for as far as 9 feet, aiming at the eyes. The venom does not affect the unbroken skin but, like that of the ringhals (*Hemachatus*), in the eyes it causes great pain and spasm of the eyelids. The eye tissues are destroyed unless the venom is washed out immediately with water or some other nonirritating liquid. Subsequent flushing of the eyes with antivenin diluted with water (1:5) apparently is beneficial since the venom is absorbed quickly into the tissues.

Polyvalent antivenins are manufactured by Behringwerke, Germany; South African Institute for Medical Research, Johannesburg, Republic of South Africa; and the Institut Pasteur, Paris.

Yellow Cobra, *Naja nivea* (Linnaeus).

Identification: A relatively small and slender cobra without the black bands under the hood which characterize the forest and spitting cobras and without the row of subocular scales that identifies the Egyptian cobra (*N. haje*). It has a broad and rather rounded hood. Adults average 5 to 6 feet; record length about 7 feet.

Dorsal coloration extremely variable, usually yellowish to reddish brown but occasionally (southern South-West Africa and adjacent Cape Province) unicolor black; light color sometimes speckled with dark, or vice-versa. Lighter, and usually unicolor, below. One

FIGURE 63.—Yellow Cobra, *Naja nivea*. This is a yellow-speckled brown individual. The hood is not fully spread. Photo by New York Zoological Society.

or two brown markings or bands below hood in young; this disappears in adults.

Dorsals in 19–21 rows at midbody, more (23) on neck, fewer (15) posteriorly. Ventrals 195–227; subcaudals 50–68.

Distribution: Temperate southern Africa, extending northward in the west to central South-West Africa; absent east of Basutoland.

Remarks: If disturbed, this cobra faces the enemy with body raised and hood expanded, ready to strike if it comes within reach. If left alone it will retreat without further signs of aggression. The venom is the most toxic of the African cobras and fatalities often result if the bite is not treated quickly.

Polyvalent antivenins are produced by the South African Institute for Medical Research.

ELAPIDAE: Genus *Paranaja* Loveridge, 1944.

Burrowing cobra.

A single, little-known species (*P. multifasciata* Werner) is known from western Central Africa. The few specimens that have been described are all small (2 feet or less in length) but have relatively large fangs. Although no bites are reported for this species, it must be regarded as a potentially dangerous snake.

Definition: Head short, flattened, slightly distinct from body. Body moderately slender, cylindrical, apparently without a hood; tail short.

Eyes of moderate size; pupils round.

Head scales: The usual 9 scales on the crown; rostral broad, rounded; internasals short. Laterally, nasal in broad contact with single preocular.

Body scales: Dorsals smooth, in 15–17 oblique rows at midbody, more (17–19) on neck, fewer (13) posteriorly. Ventrals 150–175; anal plate entire; subcaudals 30–39, all or most paired.

Maxillary teeth: Two tubular fangs with external grooves followed, after an interspace, by two small teeth.

ELAPIDAE: Genus *Pseudohaje* Günther, 1858.

Tree cobras.

Two species are recognized; both inhabit the tropical rain forest region of central and western Africa. They have average adult lengths of about 6 feet and individuals occasionally approach 8 feet. Both species are considered dangerous.

Definition: Head short and narrow, slightly distinct from neck; snout broad, rounded, canthus distinct. Body slender, tapering; neck region with very slight suggestion of hood; tail long.

Eyes very large; pupils round.

Head scales: The usual 9 on the crown; rostral broad. Laterally, nasal in contact with preocular or separated from it by a "loreal" scale that is occasionally

formed by a vertical suture across the unusually elongate preocular.

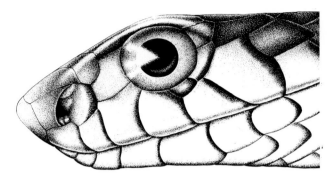

FIGURE 64.—Head Scales of *Pseudohaje*. The long preocular is sometimes broken with a vertical suture to form a "loreal." This is the only elapid snake in which this is known to occur frequently. Drawing courtesy of Charles M. Bogert.

Body scales: Dorsals smooth and glossy, in 13–15 oblique rows at midbody, the same number or more (15) on the neck, fewer (9–11) posteriorly. Ventrals 180–205; anal plate entire; subcaudals 74–94, paired.

Maxillary teeth: Two short fangs with external grooves followed, after an interspace, by 2–4 small teeth.

Gold's Tree Cobra, *Pseudohaje goldii* (Boulenger).

Identification: A long-tailed, mamba-like arboreal cobra with shiny black scales and large eyes. It (and the closely-related *P. nigra*) differs from other African cobras in the apparent absence of a hood, the few scale rows (15 at midbody), and the unusually long tail (more than 20 percent of total length, versus less than 20 percent in other cobras). It differs from the mambas

FIGURE 65.—Gold's Tree Cobra, *Pseudohaje goldii*. The large eye and the glossy scales are characteristic. Photo by New York Zoological Society. (See also plate VII, figure 2.)

97

in having a single preocular (3 in mambas) in contact with the nasal or separated from it by a "loreal" (widely separated by prefrontal in mambas). Adults average 6 to 7 feet in length; record 7 feet, 11 inches.

Dorsal surface unicolor shiny black; underneath the anterior half is yellow with black-margined ventrals, the black becoming progressively broader posteriorly, tail black.

Dorsals in 15 rows on neck and at midbody, 11 posteriorly.

Distribution: The tropical rain forest region from Nigeria eastward to Uganda and southward to Southwest Africa.

Remarks: This cobra is rarely encountered and nothing appears to be known of its venom. A closely-related and poorly-known species, *P. nigra* Günther, ranges westward to Sierra Leone.

No antivenin is produced for either of these species.

VIPERIDAE: Genus *Adenorhinos* Marx & Rabb, 1965.

Worm-eating viper.

A single species, *A. barbouri* (Loveridge), is recognized in this recently described genus from Tanzania. It is a small species, reportedly feeding on earthworms, and is not believed to be dangerous.

Definition: Head moderately broad and distinct from neck; snout short and rounded; canthus rostralis obtuse. Body moderately slender; tail moderate in length.

Eyes very large (1½ times the distance to lip); pupils vertically elliptical.

Head scales: No enlarged scutes on crown, covered with small imbricate keeled scales. Laterally, nostril in anterior part of single nasal, which has a posterior depression; nasal in contact with preocular; a single row of suboculars separating eye from upper labials; anterior and posterior temporals single.

Body scales: Dorsals keeled except for first row, in 20–23 rows at midbody. Ventrals rounded, 116–122; subcaudals single, 19–21.

Remarks: This species was recently removed from the genus *Atheris* by Marx and Rabb (1965: 186).

VIPERIDAE: Genus *Atheris* Cope, 1862.

African bush vipers.

Six species are recognized in the genus, which is restricted to tropical Africa. All of the species are relatively small, prehensile-tailed, arboreal snakes which reach a maximum length of less than 3 feet. Although few bites have been recorded, the recent description of a bite from a small specimen (Knoepffler, 1965) suggests that the bite of a 30-inch individual might be a hazard.

Definition: Head very broad and sharply distinct from narrow neck; canthus distinct, snout broad. Body relatively slender, tapering, slightly compressed; tail prehensile, moderate in length.

Eyes moderately large; pupils vertically elliptical.

Head scales: No enlarged plates on crown, covered with small imbricate scales which may be smooth or keeled. Laterally, 2–3 flat scales between nasal and eye; eye separated from labials by 1–3 rows of small scales.

Body scales: Dorsals strongly keeled, with apical pits; lateral scales smaller than those near dorsal midline, not serrated; scales in 15–36 oblique rows at midbody, fewer (11–19) posteriorly. Ventrals rounded, 142–175; subcaudals single, 38–67.

Remarks: Marx and Rabb (1965: 186) removed *A. barbouri* Loveridge from *Atheris* and made it the type of a new genus, *Adenorhinos*. (See also remarks under *Vipera superciliaris.*)

Sedge Viper, *Atheris nitschei* Tornier.

Identification: A rather stout-bodied arboreal viper, usually green with black markings. Dorsal scales relatively small and numerous. Adults average 20 to 24 inches in length; record length is 28¾ inches (Pitman, 1935: 285).

Crown of head usually green with a V-shaped or A-shaped mark; sometimes almost entirely black. Body bright to olive green, irregularly marked with black or with scales tipped with black, occasionally almost unicolor black with a lighter tail. Belly distinctly lighter than dorsal surface, yellowish or very pale green.

Dorsals in 25–32 rows at midbody, the same number or more on neck, fewer (19) posteriorly. Ventrals 143–164; subcaudals 38–58.

Distribution: Mountain areas of the eastern Congo and Uganda southward to northern Rhodesia. Sometimes found on the ground but usually in the reeds and

FIGURE 66.—Sedge Viper, *Atheris nitschei.* Photo by New York Zoological Society. (See also plate VII, figure 6.)

papyrus of lake margins or upland swamps, or in the elephant grass of humid valleys up to a height of 10 feet from the ground. Reported at altitudes of 6,000 to 7,500 feet.

Remarks: This is a very common bush viper in its rather restricted range. It appears to be a minor hazard.

No antivenin is produced for this group of vipers.

Green Bush Viper, *Atheris squamigera* (Hallowell).

Identification: A green or sometimes yellow viper without any black markings. It usually has yellow crossbands or paired yellow spots but may be almost unicolor. Dorsal scales larger and fewer than in *A. nitschei*. Body usually slender but large individuals may be quite stout. Adults average about 18 to 24 inches in length; occasional individuals approach 30 inches.

Crown of head unicolor green; labials light yellow or cream. Body green usually with 30–35 narrow yellow crossbands or paired yellow spots; sometimes unicolor green or yellow with scattered green scales. Chin yellow; belly like the dorsal surface.

Dorsals 15–23 on neck and at midbody, fewer (11–17) posteriorly. Ventrals 152–175; subcaudals 45–67.

FIGURE 67.—Green Bush Viper, *Atheris squamigera.* Photo by Isabelle Hunt Conant. (See also plate VII, figure 5.)

Distribution: The tropical rain forest region from western Kenya and the Cameroons to Angola; on the island of Fernando Po.

Remarks: This small arboreal viper, though common through the forest areas, appears to be a minor hazard. A very similar species, *A. chlorechis* (Schlegel), ranges through the forests of west Africa.

No antivenin is produced for this group of vipers.

VIPERIDAE: Genus *Atractaspis* Smith, 1849.

Mole vipers.

Sixteen species are currently recognized. All are African except for two: *A. engaddensis* Haas (which ranges from Egypt to Israel) and *A. microlepidota* (which is found in the southern part of the Arabian Peninsula as well as through much of north and central Africa). All are small snakes, less than 3 feet in length. However, they have large fangs (which look enormous in the small mouth) and are capable of inflicting serious bites on those who attempt to pick them up or who step on them with bare feet.

Definition: Head short and conical, not distinct from neck, no canthus; snout broad, flattened, often pointed. Body cylindrical, slender in small individuals, stout in larger ones; tail short, ending in distinct spine.

Eyes very small; pupils round.

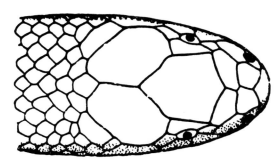

FIGURE 68.—Crown scutes of *Atractaspis irregularis.* The broad rostral and frontal, the small supraoculars, and the contact of nasal and preocular are shown. These are all characteristics of the genus. (See also plate VII, figs. 7–8.) Redrawn from Pitman, 1938.

Head scales: The usual 9 crown scales, rostral enlarged, extending between internasals to some degree, often pointed; frontal large and broad, supraoculars small. Laterally, nasal in contact with single preocular (no loreal), usually one postocular.

Body scales: Dorsals smooth without apical pits, in 19–37 nonoblique rows at midbody. Ventrals 178–370; anal plate entire or divided (the only viperid snakes with divided anal plates); subcaudals single or paired, 18–39.

Bibron's Mole Viper, *Atractaspis bibronii* Smith.

Identification: A purplish-brown or black, relatively slender viper with small head and strongly projecting snout. Adults average 15–18 inches in length; occasional individuals may slightly exceed 2 feet.

Dorsal color usually uniform, dark brown or black, often with a purplish sheen. Ventral color creamy white, yellowish, spotted with brown or entirely brown except for light anterior edges to the ventrals. Ventral color, when light, extends up onto the first 2–3 rows of dorsal scales and onto the lips.

Five supralabials and one (occasionally two) anterior temporal; third infralabial greatly enlarged, separated from its fellow below by 2–3 scales. Dorsals in 21–25 rows at midbody. Ventrals 196–260; anal plate entire; subcaudals 19–28, all or mostly single.

Distribution: These snakes usually live under stones or in burrows and are commonly seen on the surface only after heavy rains have driven them from their subterranean quarters. From Angola and southern

Rhodesia southward to southern Southwest Africa and Natal.

Remarks: This small viper is described (FitzSimons, 1962: 323) as "exceedingly irascible and is very quick to bite on the least provocation." Many people have been bitten who attempted to pick up a mole viper in the usual fashion (i.e., just behind the head). Apparently the peculiar body morphology makes this inadvisable.

In a strike, the head is thrown over the victim, the sides of the lower jaw are drawn medially to expose the fangs (often only one), and then the head is jerked downward and backward to embed the fangs in the victim.

The bite of even a small individual causes intense local pain and swelling, and often has severe systemic effects as well.

No antivenin is produced for this group of vipers.

Western Mole Viper, *Atractaspis corpulenta* (Hallowell).

Identification: A slate-colored, rather stout viper with small head and strongly-projecting snout. Adults average 18 to 20 inches in length; occasional individuals may attain a length of 2 feet.

Dorsal color slate-gray or slate-blue; often terminal portion of tail white. Lighter underneath.

Five supralabials and a single anterior temporal; second infralabial greatly enlarged, fused with chin shields and in contact with its fellow. Dorsals 23–29 at midbody. Ventrals 178–208; anal plate entire; subcaudals 22–28, all or nearly all single.

Distribution: The tropical rain forest region from the Ivory Coast to the eastern Congo.

Remarks: Little appears to have been recorded concerning the habits or the effect of the bite of this small viper. Nevertheless, it should be regarded with suspicion and treated with respect (See Remarks under *A. bibronii* and *A. microlepidota.*)

Northern Mole Viper, *Atractaspis microlepidota* Günther.

Identification: A slender, small headed and short-snouted viper that is black or dark brown above and below. Adults average 20 to 24 inches in length; occasional individuals approach a length of 30 inches.

Color uniform dark brown or black, usually with a bluish sheen, above and below.

Six or seven supralabials and 2–3 anterior temporals; none of infralabials greatly enlarged. Dorsals in 25–37 rows at midbody. Ventrals 210–245; anal plate entire or divided; subcaudals 23–37, all or mostly single.

Distribution: The savannah regions of northern and western Africa from Mauritania to Somalia, Uganda, and Kenya. It is also known from various localities in the southern part of the Arabian Peninsula (*Atractaspis microlepidota andersonii* **Boulenger**) but the relationships of this form with other species is not clear.

Remarks: This snake is one of the commonest poisonous snakes in the Sudan (Corkill, 1935: 30) and is said to be the second most frequent cause of snakebite accidents and death. Of three cases in which this mole viper was identified, one adult man died and two (also adults), though bitten by small snakes, were very seriously affected. A larger series (Corkill, 1956) gave a 25 percent mortality.

No antivenin is produced for this group of snakes.

VIPERIDAE: Genus *Bitis* Gray, 1842.

African vipers.

Ten species are found in tropical and southern Africa. They include the largest of the true vipers (Viperidae) as well as some small and moderately sized ones; all of the members of this genus are dangerous, some of them extremely so.

Definition: Head broad and very distinct from narrow neck; snout short, a distinct canthus. Body somewhat depressed, moderately to extremely stout; tail short.

Eyes small; pupils vertically elliptical.

Head scales: No enlarged plates on crown, covered with small scales. Some species have enlarged and erect scales on snout or above eye. Laterally, rostral separated from nasal by 0 (in *B. worthingtoni*) to 6 (in some *B. nasicornis*) rows of small scales, eye separated from supralabials by 2–5 rows of small scales.

Body scales: Dorsals keeled with apical pits, in 21–46 nonoblique or slightly oblique rows at midbody, fewer anteriorly and posteriorly. Ventrals rounded or with faint lateral keels, 112–153; subcaudals paired, laterally keeled in some species, 16–37.

Horned Puff Adder, *Bitis caudalis* (Smith).

Identification: A faded, light-colored desert viper with short snout and raised supraorbital ridges. A single hornlike spine over the eye is characteristic of this species, but rarely it may be absent. Similar South African vipers usually have either multiple "horns" (*B. cornuta*) or lack them entirely (*B. atropos*). Adults average 12 to 15 inches in length; record length is "close on 20 inches."

FIGURE 69.—Horned Puff Adder, *Bitis caudalis*. Photo by New York Zoological Society.

Dorsal color varies from rather dark reddish to grayish brown in the east to very light gray, buff, or pinkish in the west. A vertebral row of rectangular blotches with a usually alternating series of smaller and more rounded blotches laterally. Blotches usually with light centers and often white-edged. A dark-edged light band passes across the crown through the eyes and obliquely to the rear of the mouth; a V-shaped mark on crown. Pattern obscure in western, light-colored individuals.

Dorsals in 21–29 rows at midbody. Ventrals 120–153; subcaudals 18–34, the posterior ones usually with lateral keels.

Distribution: Desert regions of southern Angola and western Rhodesia southward through the central part of Cape Province; absent from eastern and western parts of the Cape.

Remarks: This small viper has a highly toxic venom and some deaths are reported as a result of its bite. It often conceals itself in the surface of the sand and strikes out from this position with little provocation.

Antivenins (Polyvalent, from other viper venoms) produced by Behringwerke, the Institut Pasteur and the South African Institute for Medical Research are said to be effective.

Gaboon Viper, *Bitis gabonica* (Duméril, Bibron, and Duméril).

Identification: A very large and extremely thick-bodied viper with a distinctive color pattern; crown of head tan with a narrow brown median stripe. The only snake with which it is likely to be confused is the river jack (*B. nasicornis*), which has a large and distinct arrow-shaped mark on the crown. Adults average 4 to 5 feet; record length is 6 feet 8½ inches (from Sierra Leone).

Body pattern is a complex geometrical arrangement of tans, blues, and blacks, some of the markings with white edges. The pattern may be quite brilliant but it is highly disruptive and the gaboon viper is difficult to see on the leaf-covered forest floor.

A pair of triangular nasal "horns," much more evident

Figure 70.—Gaboon Viper, *Bitis gabonica.* Photo by Charles Hackenbrock and Staten Island Zoo.

in west African individuals. Dorsals in 28–46 slightly oblique rows at midbody. Ventrals 125–140; subcaudals 17–33.

Distribution: Tropical rain forests and their immediate environs from Sierra Leone and southern Sudan southward to northern Angola and northern Natal.

Remarks: This is the largest of the vipers. The fangs are almost 2 inches long in large individuals and there are very large amounts of highly toxic venom. However, the gaboon viper is nocturnal in habit and difficult to awaken in the daytime. Relatively few bites are inflicted by this sluggish snake but they are very serious and usually are lethal without prompt treatment.

Polyvalent antivenins are produced by Behringwerke, the Institut Pasteur and the South African Institute for Medical Research.

Puff Adder, *Bitis arietans* (Merrem).

Identification: The rough-scaled appearance and the alternating pattern of dark and light chevron-shaped

Figure 71.—Puff Adder, *Bitis arietans.* Photo by Zoological Society of San Diego. (See also plate II, figure 4.)

markings are characteristic. Head lanceolate; nostrils face more directly upward than in other African vipers. Adults average 3 to 4 feet; occasional individuals attain a length of 5 feet.

A light band crosses head between eyes and is continued as a diagonal band from the eye to the rear of the mouth. Ground color varies from light grayish tan or yellow to dark brown; either the light or the dark chevron series may be emphasized, depending on the density of the ground color.

Dorsals in 29–41 rows at midbody. Ventrals 124–147; subcaudals 16–37.

Distribution: Savannah and grasslands from Morocco and western Arabia throughout Africa except for the Sahara and rain forest regions. Found from sea level to elevations of at least 9,000 feet.

Remarks: Due to its wide distribution, common occurrence, and lethal potential, the puff adder probably kills more people than any other African snake. The venom is strongly toxic, causing widespread tissue destruction and bleeding from mucous membranes, both externally and internally. A puff adder of average dimensions may have sufficient venom to kill 4 to 5 men. Death may not occur for more than 24 hours and is usually preceded by severe internal hemmorhages. The snake is nocturnal, slow moving, and tends to remain immobile when approached.

Antivenins are produced by Behringwerke, the Institut Pasteur, and the South African Institute for Medical Research.

River Jack, *Bitis nasicornis* (Shaw).

Identification: A large and extremely thick-bodied viper with relatively small head and two or three pairs of nasal "horns." Most easily distinguished from the Gaboon viper (which may have a pair of nasal "horns") by the large dark arrow-shaped mark on the crown. Adult snakes are 2½ to 3½ feet in length; exceptional individuals attain a length of 4 feet.

Body pattern very complex, usually made up of a vertebral series of 15-18 paired, yellow-edged blue blotches, with a lateral series of light-edged dark triangles extending up from the belly. Ground color varies through various shades of blue, pink, purple, and green. In spite of its brilliant colors the pattern blends well with the forest floor.

Dorsals in 35–41 rows at midbody. Ventrals 124–140; subcaudals 16–32.

FIGURE 72.—River Jack, *Bitis nasicornis*. The arrow-shaped head marking is distinctive. Photo by Isabelle Hunt Conant. (See also plate II, figure 3.)

Distribution: Swamps, river banks, and other moist habitats through the tropical rain forest region from Liberia and Uganda southward through the Congo region.

Remarks: The river jack has a more restricted range than the gaboon viper and apparently inflicts even fewer bites. However, its venom is reported to be highly toxic and it is not as placid as the latter.

Antivenins are produced by Behringwerke, the Institut Pasteur, and the South African Institute for Medical Research.

VIPERIDAE: Genus *Causus* Wagler, 1830.

Night adders.

Four species are found in tropical and southern Africa. None attains a length of over 3 feet. The fangs are relatively small, and the venom is rather mildly toxic. They look surprisingly like nonpoisonous snakes. Night adders are not considered dangerous to life but their bite is painful.

Definition: Head moderate in size, fairly distinct from neck, an obtuse canthus. Body cylindrical or slightly depressed, moderately slender; tail short.

Eyes moderate in size; pupils round.

Head scales: The usual 9 crown scales; rostral broad, sometimes pointed and upturned; frontal long, supraoculars large. Laterally, a loreal present, separating nasal and preoculars; suboculars present, separating eye from labials.

Body scales: Dorsals smooth or weakly keeled, with apical pits, in 15–22 oblique rows at midbody, fewer (11–14) posteriorly. Ventrals rounded, 109–155; subcaudals single or paired, 10–33.

Rhombic Night Adder, *Causus rhombeatus* (Lichtenstein).

Identification: A satiny sheen to the scales and a V-shaped marking on the back of the head are characteristic of this snake. It differs from the other night adders in having a rounded snout and a relatively unmodified rostral scute. Adults average about 2 feet in length; exceptional individuals reach "close on 3 feet."

Ground color light gray to dark brown or olive with a series of 20–30 squarish blotches down the back; irregular markings laterally. Markings are often white-edged. Unicolor white or yellowish below, ventrals occasionally have dark edges.

Distribution: Widely distributed through the savan-

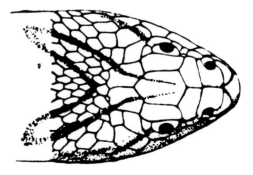

FIGURE 73.—Head scales of Rhombic Night Adder, *Causus rhombeatus*. Note the rear projection of the rostral, the presence of a loreal, and of suboculars; all of these are characteristic of the genus. (See also plate VIII, figs. 4–6). Redrawn from Pitman, 1938.

nah and grasslands, but preferring damp areas. From Sudan and Somalia to Angola in the west, and along the eastern Cape to its tip.

Remarks: This nocturnal viper is usually inoffensive and most bites are a result of persons stepping on the snake in the dark. The venom is not of high toxicity and a bite usually results in nothing more than local swelling and pain. If teased, the adder flattens the neck, puffs up the body and hisses loudly, striking out wildly at any moving object. No antivenin is produced for this group of vipers.

VIPERIDAE: Genus *Echis* Merrem, 1820.

Saw-scaled vipers.

Two species are recognized. One (*E. coloratus*) is restricted to eastern Egypt, the Arabian Peninsula, and Israel. The other (*E. carinatus*) ranges from Ceylon and southern India across western Asia and north Africa southward into tropical Africa. Although neither attains a length of 3 feet, they possess a highly toxic venom and are responsible for many deaths. When disturbed they characteristically inflate the body and produce a hissing sound by rubbing the saw-edged lateral scales against one another. This same pattern of behavior is shown by the nonpoisonous egg-eating snakes *Dasypeltis*.

Definition: Head broad, very distinct from narrow neck; canthus indistinct. Body cylindrical, moderately slender; tail short. (See p. 83 and fig. 52).

Eyes moderate in size; pupils vertically elliptical.

Head scales: A narrow supraocular sometimes present; otherwise crown covered with small scales, which may be smooth or keeled. Rostral and nasals distinct. Laterally eye separated from labials by 1–4 rows of small scales; nasal in contact with rostral or separated from it by a row of small scales.

Body scales: Dorsals keeled, with apical pits, lateral scales smaller, with serrate keels, in 27–37 oblique rows at midbody. Ventrals rounded, 132–205; subcaudals single, 21–52.

VIPERIDAE: Genus *Vipera* Laurenti, 1768.

True adders.

Eleven species are recognized. This is an especially variable group, with some members that are small and relatively innocuous (e.g., *V. berus*) and others that are extremely dangerous (*V. lebetina*, *V. russelii*). They are found from northern Eurasia throughout that continent and into north Africa. One species ranges into the East Indies (*V. russelii*), and two are found in east Africa (see Remarks under *V. superciliaris*).

Definition: Head broad, distinct from narrow neck; canthus distinct. Body cylindrical, varying from moderately slender to stout; tail short.

Eyes moderate in size to small; pupils vertically elliptical.

Head scales: Variable: one species (*V. ursinii*) has all 9 crown scutes, most species have at least the supra-

oculars, but even these are absent in one (*V. lebetina*): head otherwise covered with small scales. Laterally, nasal in contact with rostral or separated by a single enlarged scale (the nasorostral), eye separated from supralabials by 1–4 rows of small scales.

Body scales: Dorsals keeled, with apical pits, in 19–31 nonoblique rows at midbody. Ventrals rounded, 120–180; subcaudals paired, 20–64.

African Lowland Viper, *Vipera superciliaris* Peters.

Identification: The only viper in central Africa with supraocular plates but with the remainder of the crown covered with small scales. Adults average about 20 to 24 inches in length.

Eyes moderate in size; snout rounded with distinct canthus.

Head with symmetrical black markings; body pale reddish-brown with a vertebral row of black crossbars which are broken laterally by an interrupted yellowish stripe. Ventral surface white with black spots.

Head covered with small imbricate keeled scales except for the large supraoculars. Dorsals in 26–30 rows at midbody. Ventrals 142–159; subcaudals 32–43.

Distribution: Lowland areas of Tanzania (near Lake Nyasa) and Mozambique.

Remarks: No bites by this snake have been reported. It is a rare snake in collections and its relationships (together with that of *V. hindii* Boulenger) are still in dispute. Kramer (1961) believes that these two vipers are closely related to the desert-dwelling members of the genus *Bitis* (e.g., *B. caudalis*), while Marx and Rabb (1965) believe that they represent a terrestrial branch of the genus *Atheris*.

REFERENCES

(*See also General References*)

ANGEL, F. 1933. Les serpents de l'Afrique occidentale Francaise. Larose Ed., Paris. 246 p., 83 figs.

BOGERT, Charles M. 1940. Herpetological Results of the Vernay Angola Expedition, with Notes on African Reptiles in Other Collections. Part I. Snakes, Including an Arrangement of African Colubridae. Bull. Amer. Mus. Nat. Hist., 77 (Art 1): 1–107, figs. 1–18, pl. 1.

BOGERT, Charles M. 1942. *Pseudohaje* Gunther, A Valid Genus for Two West African Arboreal Cobras. Amer. Mus. Novitates (1174): 1–9, figs. 1–8.

BROADLEY, Donald G. 1959. The Herpetology of Southern Rhodesia. Part 1. Snakes. Bull. Mus. Comp. Zool., 120 (1): 1–100, figs. 1–10, pls. 1–6.

REFERENCES (continued)

CANSDALE, George S. 1961. West African Snakes. Longmans Green & Co., Ltd. London. 74 p., 34 figs. (most in color).

CORKILL, N. L. 1935. Notes on Sudan Snakes. Publ. Sudan Govt. Mus. (Nat. Hist.), (3) : (*not seen*).

CORKILL, Norman L. 1956. Snake Poisoning in the Sudan, p. 331–339, figs. 1–5. *In* E. E. Buckley and N. Porges, Venoms, Pub. Am. Assoc. Advanc. Sci. (44). 467 p.

DOUCET, Jean. 1963. Les serpents de la Republique de Cote d'Ivoire. Acta Tropica, 20 (3 & 4) : 201–340, figs. 1–57, pls. 1–10.

FITZSIMONS, Vivian F. M. 1962. Snakes of Southern Africa. Macdonald & Co., Ltd., London. 423 p., 74 color pls., 106 figs., 78 maps.

KNOEPFFLER, Louis-Philippe. 1965. Auto-observation par morsure d'*Atheris* sp. Toxicon, 2 : 275–276.

KRAMER, Eugen. 1961. Uber zwei afrikanische Zwergpuffottern, *Bitis hindii* (Boulenger, 1910) und *Bitis superciliaris* Peters, 1854). Vierteljahrsschrift naturf. Ges. Zurich, 106 : 419–423, fig. 1.

LAURENT, Raymond F. 1956. Contribution a l'Herpetologie de la region des Grands Lacs de l'Afrique central. Parts 1–3. Ann. Mus. Royal Congo Belgie, ser. 8 (sci. zool.), 48 : 1–390, figs. 1–50, pls. 1–31.

LAURENT, Raymond F. 1956. Esquisse d'une faune herpetologique du Ruanda-Urundi. Bull. Nat. Belges, nov-dec. 1956 : 280–287.

LAURENT, Raymond F. 1964. Reptiles et Amphibiens de l'Angola. Pub. Culturais Mus. Dundo (67) : 1–165, figs. 1–40.

LEESON, Frank. 1950. Identification of Snakes of the Gold Coast. Crown Agents for the Colonies, London. 142 p., 65 figs., 33 pls. (13 color).

LOVERIDGE, Arthur. 1953. Zoological Results of a Fifth Expedition to East Africa. III. Reptiles from Nyasaland and Tete. Bull. Mus. Comp. Zool., 110 (3) : 143–322, figs. 1–4, pls. 1–5.

LOVERIDGE, Arthur. 1957. Check List of the Reptiles and Amphibians of East Africa (Uganda; Kenya; Tanganyika; Zanzibar). Bull. Mus. Comp. Zool., 117 (2) : 153–362.

MANACAS, Sara. 1956. Ofidios de Mocambique. Mem. Junta Invest. Ultram., 8 : 135–160.

MERTENS, Robert. 1938. Herpetologische Ergebnisse einer Reise nach Kamerun. Abh. senckenbergischen naturf. Ges., 442 : 1–52, pls. 1–10.

MERTENS, Robert. 1955. Die Amphibien und Reptilien Sudwestafrikas. Aus den Ergebmissen einer im Jahre 1952 ausgefuhrten Reise. Abhandl. senckenbergischen naturf. Ges., 490 : 1–172, 24 pls. (1 color).

PARKER, H. W. 1949. The Snakes of Somaliland and the Sokotra Islands. Zool. Verh. Rijksmus. Nat. Hist. Leiden (6) : 1–115, figs. 1–11, map.

PERRET, J. L. 1960. Une nouvelle et remarquable espece d'Atractaspis (Viperidae) et quelques autres Serpents d'Afrique. Rev. Suisse Zool., 67 (5) : 129–139, figs. 1-4.

PITMAN, Charles R. S. 1938. A Guide to the Snakes of Uganda. Pub. Uganda Soc., Kampala, Uganda. 362 p., 2 figs., 23 color pls.

POPE, Clifford H. 1958. Fatal Bite of Captive African Rear-fanged Snake (*Dispholidus*). Copeia, 1958 (4) : 280–282.

SCHMIDT, Karl P. 1923. Contributions to the Herpetology of the Belgian Congo Based on the Collection of the American Museum Congo Expedition, 1909–1915. Part II. Snakes. Bull. Amer. Mus. Nat. Hist., 49 (art. 1) : 1–146, figs. 1–15, pls. 1–22, maps 1–19.

SWEENEY, R. C. H. 1961. Snakes of Nyasaland. Nyasaland Soc. & Nyasaland Govt. Zomba, Nyasaland. 200 p., 43 figs., map.

VISSER, John. 1966. Poisonous Snakes of Southern Africa and the Treatment of Snakebite. Howard Timmins : Capetown, 60 pp., 65 figs. (60 col.).

WITTE, Gaston-Francois de. 1941. Batraciens et reptiles. *In* Exploration du Parc National Albert, Mission G. F. de Witte (1933–1935). Pub. Inst. Parc Nationaux Congo Belge, fasc. 33 : 1–261, figs. 1–54, pls. 1–76.

WITTE, Gaston-Francois de. 1962. Genera des serpents du Congo et du Ruanda-Urundi. Ann. Mus. Royal Afrique Central, ser. 8 (104) : 1–203, figs. 1–94, pls. 1–15.

THE NEAR AND MIDDLE EAST

Definition of the Region:

Includes Asian Turkey, Syria, Jordan, Israel, Lebanon, the Arabian Peninsula, Iraq, Iran, Afghanistan and the Georgian, Armenian Azerbaidzhan, Turkmen, Uzbek, Tadzhik, Kirgiz and Kazakh Socialist Soviet Republics.

TABLE OF CONTENTS

TABLE 11.—DISTRIBUTION OF POISONOUS SNAKES IN THE NEAR & MIDDLE EAST*

	Asian Turkey	Syria	Jordan	Israel & Lebanon	Iraq & Kuwait	Saudi Arabia	Yemen & Aden	Muscat & Oman	Iran	Afghanistan	Georgian, Armenian & Azerbaidzhan S.S.R.	Turkmen & Uzbek S.S.R.	Tadzhik & Kirgiz S.S.R.	Kazakh S.S.R.
ELAPIDAE														
Naja haje		S	E			W	X							
Naja naja									NE	X		S	SW	
Walterinnesia aegyptia			X	X	X	N			W					
VIPERIDAE														
Atractaspis engaddensis			W	S										
A. microlepidota							X							
Bitis arietans						SW	X							
Cerastes cerastes		?	X	S	W	X								
C. vipera				S		?								
Echis carinatus					S	X	X	X	X	X				
E. coloratus			X	S		X	X	?						
Eristicophis macmahonii									SE	SW				
Pseudocerastes persicus		?	X	S	W	N		?	X	S		S	W	
Vipera ammodytes	X	NW							NW					
V. berus	NE											W		N
V. kaznakovi									NW		X			
V. lebetina	X	X	?	N	X				X	X	X	S	SW	S
V. ursinii	?								NW		X	SW	W	X
V. xanthina	W	X	X	X					N		X			
CROTALIDAE														
Agkistrodon halys									N	?	N	X	X	X

*Certain groups of adjoining nations or provinces are here treated as units. The symbol X indicates distribution of the species is widespread within the unit. Restriction of a species to part of a unit is indicated appropriately (SW = southwest, etc.). The symbol ? indicates suspected occurrence of the species within the unit without valid literature.

INTRODUCTION

Like north Africa this is a predominantly arid region, although it does not contain quite so much sterile desert. This trend toward a drier climate is quite recent, marked changes having occurred within historic times. Overgrazing, deforestation, and other forms of human misuse have contributed to the trend. The snake fauna contains species in common with northern Africa, Europe, and central Asia; toward the east there is infiltration of species characteristic of tropical Asia.

In the Middle East also, the vipers cause most of the snakebites. Cobras and other elapids occur, but are rare or restricted in range, and inflict few bites. Several species of sea snakes are encountered in the Persian Gulf.

GENERIC AND SPECIES DESCRIPTIONS

ELAPIDAE: Genus *Naja* Laurenti, 1768.

Cobras.

Six species are recognized; all are African except the Asiatic cobra, *Naja naja*, and range throughout the African continent except for the drifting sand areas of the Sahara region. They are snakes of moderate (4 feet) to large (8 feet) size, with large fangs and toxic venom. The *N. nigricollis* species "spits" its venom at the eyes of an aggressor; it is found in the southern part of the region of north Africa. The Egyptian cobra (*Naja haje*) and the western subspecies of the Asiatic cobra (*Naja naja oxiana*) are found in the Near and Middle East region (see p. 80 and p. 124).

Definition: Head rather broad, flattened, only slightly distinct from neck; snout rounded, a distinct canthus. Body moderately slender, slightly depressed, tapered; neck capable of expansion into hood; tail of moderate length.

Eyes moderate in size; pupils round.

Head scales: The usual 9 on the crown; frontal short; rostral rounded. Laterally, nasal in contact with the one or two preoculars.

Body scales: Dorsals smooth, in 17–25 oblique rows at midbody, usually more on the neck, fewer posteriorly. Ventrals 159–232; anal plate entire; subcaudals 42–88, mostly paired.

Maxillary teeth: Two rather large tubular fangs with external grooves followed, after an interspace, by 0–3 small teeth.

ELAPIDAE: Genus *Walterinnesia* Lataste, 1887.

Desert black snake.

A single species, *W. aegyptia*, is known from the desert regions of Egypt to Iran. It is relatively large, 3 to 4 feet, and is probably a dangerous species (see p. 81 and fig. 50).

Definition: Head relatively broad, flattened, distinct from neck; snout broad, a distinct canthus. Body cylindrical and tapered, moderately slender; tail short.

Eyes moderate in size; pupils round.

Head scales: The usual 9 on the crown; rostral broad. Laterally, nasal in contact with single elongate preocular.

Body scales: Dorsals smooth at midbody, feebly keeled posteriorly, in 23 rows at midbody, more (27) anteriorly. Ventrals 189–197; anal plate divided; subcaudals 45–48, first 2–8 single, remainder paired.

Maxillary teeth: Two large tubular fangs with external grooves followed, after an interspace, by 0–2 small teeth.

MAP 8.—Section 7, Near and Middle East.

VIPERIDAE: Genus *Atractaspis* Smith, 1849.

Mole vipers.

Sixteen species are currently recognized. All are African except for *A. engaddensis* Haas (which ranges from Egypt to Israel) and *A. microlepidota* (which is found in the southern part of the Arabian Peninsula as well as through much of north and central Africa). All are small snakes, less than 3 feet in length. However, they have large fangs (which look enormous in the

KEY TO GENERA

1. A. Crown of head with 7 or 9 large shields (see
      fig. 6) _____ 2
   B. Crown of head with small scales (see fig. 96), or
      with 6 or fewer shields_____ 7
2. A. Loreal pit present _____ *Agkistrodon*
   B. Loreal pit absent_____ 3
3. A. All subcaudals single; eye very small_____ *Atractaspis*
   B. At least some of the subcaudals paired; eye large
      to very small_____ 4
4. A. Loreal present_____ NP*
   B. Loreal absent_____ 5
5. A. Body scales smooth anteriorly, keeled posteriorly;
      anal plate divided; some subcaudals single_____ *Walterinnesia*
   B. Without the above combination of characters_____ 6
6. A. Hood present in life; body scales smooth; anal
      plate single_____ *Naja*
   B. No hood; body scales smooth or keeled; anal plate
      divided or single_____ NP
7. A. Ventrals extend full width of belly_____ 8
   B. Ventrals do not extend full width of belly_____ NP
8. A. Subcaudals single_____ *Echis*
   B. Subcaudals paired_____ 9
9. A. Horn-like process covered with small scales above
      eye_____ *Pseudocerastes*
   B. Horn-like process absent or composed of a single
      scale or spine_____ 10
10. A. Ventrals with lateral ridge or keel_____ 11
    B. Ventrals without lateral ridge or keel_____ 12
11. A. Horn-like spine often present above eye; rostral
       not flanked by enlarged scales_____ *Cerastes*
    B. No horn-like projection above eye; rostral flanked
       by enlarged scales_____ *Eristicophis*
12. A. Nostril directed upward; 27 or more scale rows
       around midbody _____ *Bitis*
    B. Nostril directed laterally; fewer than 27 scale rows
       around midbody_____ *Vipera*

_____
* NP = Nonpoisonous

small mouth) and are capable of inflicting serious bites on those who attempt to pick them up or who step on them with bare feet.

Definition: Head short and conical, not distinct from neck, no canthus; snout broad, flattened, often pointed. Body cylindrical, slender in small individuals, stout in larger ones; tail short, ending in a distinct spine.

Eyes very small; pupils round.

Head scales: The usual 9 crown scales, rostral enlarged, extending between internasals to some degree, often pointed; frontal large and broad, supraoculars small. Laterally, nasal in contact with single preocular (no loreal), usually one postocular.

Body scales: Dorsals smooth without apical pits, in 19–37 nonoblique rows at midbody. Ventrals 178–370; anal plate entire or divided (the only viperid snakes with divided anal plates); subcaudals single or paired, 18–39.

Middle East Mole Vipers, *Atractaspis*.

Identification: This group of very distinctive snakes has two representatives in the region, the Arabian mole viper (*A. microlepidota andersoni*) and oasis mole viper (*A. engaddensis*). They are very similar in appearance. Smooth scales, small head not distinct from the neck, and the typical 9 crown shields distinguish them from other vipers of the region. Large ventrals extending the width of the belly distinguish them from the burrowing blind snakes and sand boas. The pointed snout, tiny eyes (diameter less than half the distance from eye to lip), unpaired subcaudals, and overall blackish color distinguish them from the small colubrids of the Middle East. The elongated tubular venom glands extend through the anterior fifth of the body.

The Arabian mole viper has 23 or 25 scale rows at midbody, 219–254 ventrals, and 31 or fewer subcaudals. The oasis mole viper has 27 to 29 scale rows at midbody, 264–282 ventrals, and 36 to 39 subcaudals.

The average length of both species is 18 to 25 inches.

Distribution: The Arabian mole viper occurs in the southwestern part of the Arabian Peninsula. The oasis mole viper is found in Israel, Sinai and northeastern Egypt.

Remarks: These are nocturnal burrowing snakes found mainly in oases and in cultivated areas rather than in desert. Authorities who know these snakes in life agree on one point—it is impossible to hold a mole viper safely except with forceps or tongs. The small head, very flexible neck, long fangs and extraordinary ability to use one fang at a time with the jaws almost closed, make them very hazardous to handle. Several of the bites reported have been inflicted on zoologists or others collecting snakes. Most bites have been more uncomfortable than alarming, but there have been enough fatalities recorded to confirm the dangerous nature of these snakes. Local pain and swelling are seen regularly with mole viper bites. Severe cases show fever, vomiting, and blood in the urine. There is no antivenin.

VIPERIDAE: Genus *Bitis* Gray, 1842.

African vipers.

Ten species are found in tropical and southern Africa. They include the largest of the true vipers (Viperidae) as well as some small and moderately sized ones; all of the members of this genus are dangerous, some of them extremely so. The puff adder, *B. arietans*, is the only member of this group that enters the region (see p. 101).

Definition: Head broad and very distinct from narrow neck; snout short, a distinct canthus. Body somewhat depressed, moderately to extremely stout; tail short.

Eyes small; pupils vertically elliptical.

Head scales: No enlarged plates on crown, covered with small scales. Some species have enlarged and erect scales on snout or above eye. Laterally, rostral separated from nasal by 0 (in *B. worthingtoni*) to 6 (in some *B. nasicornis*) rows of small scales, eye separated from supralabials by 2–5 rows of small scales.

Body scales: Dorsals keeled with apical pits, in 21–46 nonoblique or slightly oblique rows at midbody, fewer anteriorly and posteriorly. Ventrals rounded or with faint lateral keels, 112–153; subcaudals paired, laterally keeled in some species, 16–37.

FIGURE 74.—Desert Horned Viper, *Cerastes cerastes*. A vicious but not especially dangerous snake. Photo by Standard Oil Company.

VIPERIDAE: Genus *Cerastes* Laurenti, 1768.

Horned vipers.

Two species are recognized; both are restricted to the desert regions of northern Africa and western Asia. Neither is a large species; the bite is painful but usually not serious. Both species are found in this region (see p. 82).

Definition: Head broad, flattened, very distinct from neck; snout very short and broad, canthus indistinct. Body depressed, tapered, moderately slender to stout; tail short.

Eyes small to moderate in size; pupils vertically elliptical.

Head scales: Head covered with small irregular, tubercularly-keeled scales; a large erect, ribbed horn-like scale often present above the eye; no other enlarged scales on crown. Laterally, nasal separated from rostral by 1–3 rows of small scales; eye separated from supralabials by 3–5 rows of small scales.

Body scales: Dorsals with apical pits, large and heavily keeled on back, smaller laterally, oblique, with serrated keels, in 23–35 rows at midbody. Ventrals with lateral keel, 102–165; subcaudals keeled posteriorly, all paired, 18–42.

VIPERIDAE: Genus *Echis* Merrem, 1820.

Saw-scaled vipers.

Two species are recognized. One (*E. coloratus*) is restricted to eastern Egypt, the Arabian Peninsula and Israel. The other (*E. carinatus*) ranges from Ceylon and southern India across western Asia and north Africa southward into tropical Africa. Although neither attains a length of 3 feet, they possess a highly toxic venom and are responsible for many deaths. When disturbed they charcteristically inflate the body and produce a hissing sound by rubbing the saw-edged lateral scales against one another. This same pattern of behavior is shown by the nonpoisonous egg-eating snakes *Dasypeltis* (see p. 83 and fig. 52).

Definition: Head broad, very distinct from narrow neck; canthus indistinct. Body cylindrical, moderately slender; tail short.

Eyes moderate in size; pupils vertically elliptical.

Head scales: A narrow supraocular sometimes present; otherwise crown covered with small scales, which may be smooth or keeled. Rostral and nasals distinct. Laterally eye separated from labials by 1–4 rows of small scales; nasal in contact with rostral or separated from it by a row of small scales.

Body scales: Dorsals keeled, with apical pits, lateral scales smaller, with serrate keels, in 27–37 oblique rows at midbody. Ventrals rounded, 132–205; subcaudals single, 21–52.

VIPERIDAE: Genus *Eristicophis* Alcock and Finn, 1897.

Asian sand viper.

A single species, *E. macmahonii* Alcock and Finn, is known from the desert areas of southeastern Iran, Afghanistan, and West Pakistan. It is a rather small

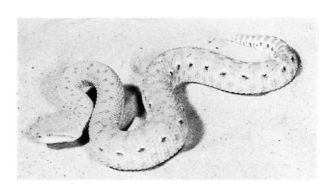

FIGURE 75.—Asian Sand Viper, *Eristicophis macmahonii.* A little-known desert viper inhabiting sand dunes. Photo by New York Zoological Society.

snake, less than 3 feet in length. However, fatal cases attributed to this species (Shaw, 1925) and a recent serious bite indicate that it is a dangerous snake with venom similar to that of *Echis*.

Definition: Head broad and flattened, very distinct from neck; snout broad and short, canthus not distinct. Body slightly depressed, moderately to markedly stout; tail short.

Eyes moderate in size; pupils vertically elliptical.

Head scales: Crown covered by small scales; rostral broad, bordered dorsally and laterally by greatly enlarged nasorostral scales. Laterally, eye separated from labials by 3–4 rows of small scales; nasal separated from rostral by nasorostral scale.

Body scales: Dorsals keeled, short, in 23–26 vertical rows at midbody. Ventrals with lateral keels, 140–148; subcaudals paired, without keels, 29–36.

VIPERIDAE: Genus *Pseudocerastes* Boulenger, 1896.

False horned viper.

A single species is recognized (see Remarks). It ranges from Sinai and the Arabian Peninsula eastward to West Pakistan. It attains a length of 3 feet and is considered dangerous.

Definition: Head broad, very distinct from neck; snout short and broadly rounded; nostrils dorsolateral, valves present.

Eyes small to moderate; pupils vertically elliptical.

Head scales: Crown covered with small imbricate scales; an erect hornlike projection covered with imbricate scales above eye. Laterally, nasals separated from rostral by small scales; eye separated from labials by 3–4 rows of small scales.

Body scales: Dorsals weakly to moderately keeled, in 21–25 nonoblique rows at midbody. Ventrals 134–158; subcaudals paired, 35–48.

Remarks: Four species were listed in this genus by Klemmer (1963: 377). However, Anderson (1963: 472) reported that *P. latirostris* Guibe, 1957, was a misidentified specimen of *Eristichophis macmahonii* Alcock and Finn, 1897. Marx and Rabb (1965: 167–175) corroborated this allocation. They also referred *P. bicornis* Wall, 1913, to the synonymy of *P. persicus* (Duméril, Bibron, and Duméril), 1954, and accorded *P. fieldii* Schmidt, 1930, subspecific status. The remaining species, *P. persicus*, was allocated to the genus *Vipera*.

Persian Horned Viper, *Pseudocerastes persicus* (Duméril, Bibron, and Duméril).

Identification: Differs from the African horned desert

FIGURE 76.—Persian Horned Viper, *Pseudocerastes persicus.* Photo by Hymen Marx.

viper in absence of keels on the ventrals and in the nature of the horn which is composed of several small scales in this species and a single spinelike scale in the African species. From other vipers within its range, it differs in the presence of the supraocular horns and dorsolateral position of the nostril.

Color pale gray or bluish gray to khaki with darker blotches or crossbands; dark band on side of head; belly white.

Average adult length 22 to 28 inches; maximum about 35 inches.

Distribution: From the Sinai Peninsula eastward through Baluchistan. Found in sandy and rocky country to elevations of about 6,000 feet; frequents burrows and crevices among rocks.

Remarks: Almost exclusively nocturnal. Hisses loudly when disturbed but is not particularly vicious. Venom of the subspecies, *P.p. fieldii* is highly toxic but produces little local tissue damage. It should be considered a dangerous snake. Antivenin is produced by the State Razi Institute, Tehran, Iran.

VIPERIDAE: Genus *Vipera* Laurenti, 1768.

True adders.

Eleven species are recognized. This is an especially variable group, with some members that are small and relatively innocuous (e.g., *V. berus*) and others that are extremely dangerous (*V. lebetina, V. russelii*). They are found from northern Eurasia throughout that continent and into north Africa. One species ranges into the East Indies (*V. russelii*), and two are found in east Africa (see Remarks under *V. superciliaris*). The long-nosed sand viper, *V. ammodytes*, and the European viper, *V. berus*, enter this region from the north and west (see p. 74 and p. 75).

Definition: Head broad, distinct from narrow neck; canthus distinct. Body cylindrical, varying from moderately slender to stout; tail short.

Eyes moderate in size to small; pupils vertically elliptical.

Head scales: Variable: one species (*V. ursinii*) has all 9 crown scutes, most species have at least the supraoculars, but even these are absent in one (*V. lebetina*): head otherwise covered with small scales. Laterally, nasal in contact with rostral or separated by a single enlarged scale (the nasorostral), eye separated from supralabials by 1–4 rows of small scales.

Body scales: Dorsals keeled, with apical pits, in 19–31 nonoblique rows at midbody. Ventrals rounded, 120–180; subcaudals paired, 20–64.

Levantine Viper, *Vipera lebetina* (Linnaeus).

Identification: Head triangular, rather long, distinct from neck; body stout; tail tapers abruptly behind vent. Crown with small keeled scales; nostril lateral; supra-

FIGURE 77.—Levantine Viper, *Vipera lebetina.* Photo by New York Zoological Society.

ocular divided into 3 small shields; usually 3 scale rows between eye and upper labials; dorsal body scales keeled, in 23 to 27 rows at midbody; ventrals not keeled; subcaudals divided.

Dorsal color light gray, khaki, or buff with minute darker punctation giving a generally dusty appearance; series of small, rectangular brown, reddish or gray blotches; belly buff variably clouded with gray; tail pinkish brown. A reddish brown phase without blotches is seen in parts of the range; in other areas the snakes may be almost uniformly dusty gray or khaki.

Average length 30 to 45 inches; maximum a little over 5 feet.

Distribution: Cyprus and the Cyclades Islands through the Caucasus and Middle East to Kashmir. Inhabits barren rocky areas usually at altitudes of 3,000 to 7,000 feet but at lower altitudes toward the northern and western part of the range.

Remarks: Very slow to move; seeming almost oblivious to stimuli when encountered by day. (Many of its local names mean "deaf one" or "blind one.") More active and alert at night but may strike quickly and savagely at any time; occasionally climbs into bushes.

The Levantine viper is important as a cause of snakebite in the Middle East. The quantity and toxicity of the venom are about the same as for Russell's viper. Antivenins against *V. lebetina* venom are produced by the State Razi Institute, Tehran, Iran, and Tashkent Institute, Moscow.

V. lebetina, like the majority of vipers found in the Middle East, lays eggs.

Near East Viper, *Vipera xanthina* (Gray).

Identification: Head large, a little shorter than in the Levantine viper; body build similar; supraocular not divided, narrow, turned up into a hornlike process in the subspecies *raddei;* usually 1 or 2 scale rows be-

tween eye and upper labials; dorsal scales usually in 23, less often in 25 rows at midbody.

Ground color sandy yellow, golden brown, gray or reddish brown with series of oval or round spots with lighter centers and pale edges; these are often fused into a zigzag band. Top of head with conspicuous V-shaped dark mark or pair of elongate dark spots; prominent dark stripe behind eye; belly yellowish with black or gray mottling.

Average length 28 to 38 inches; maximum about 4 feet.

Distribution: The Caucasus Mountains and northwestern Iran, western Turkey and south to Israel and Jordan. Occurs along stream valleys and in other places where there are vegetation and moisture; absent from true desert; often plentiful in cultivated regions.

Remarks: A nocturnal snake that not infrequently may be found near human habitation. It is alert and strikes quickly when disturbed. The subspecies *palaestinae* is the leading cause of snakebite accidents in Israel and adjoining territory.

Much research has been done on the venom of the Palestine viper (*V.x. palaestinae*). The lethal dose for man is estimated at 75 mg.—well within the capacity of an adult snake. The case fatality rate is about 5 percent. Antivenin is produced by the Institut Pasteur, Paris. An Israeli antivenin is reported to be ready for production.

CROTALIDAE: Genus *Agkistrodon* Beauvois, 1799.

Moccasins and Asian pit vipers.

Twelve species are recognized. Three of these are in North and Central America; the others are in Asia, with one species, *A. halys* (Pallas) ranging westward to southeastern Europe. The American copperhead (*A. contortrix*) and the Eurasian mamushi and its relatives (*A. halys*) seldom inflict a serious bite but *A. acutus* and *A. rhodostoma* of southeastern Asia, as well as the cottonmouth (*A. piscivorus*) of the southeastern United States, are dangerous species. Pallas's viper, *A. halys* is the only one that enters this region (see p. 75).

Definition: Head broad, flattened, very distinct from narrow neck; a sharply-distinguished canthus. Body cylindrical or depressed, tapered, moderately stout to stout; tail short to moderately long.

Eyes moderate in size; pupils vertically elliptical.

Head scales: The usual 9 on the crown in most species; internasals and prefrontals broken up into small scales in some Asian forms; a pointed nasal appendage in some. Laterally, loreal pit separated from labials or its anterior border formed by second supralabial. Loreal scale present or absent.

Body scales: Dorsals smooth (in *A. rhodostoma* only) or keeled, with apical pits, in 17–27 nonoblique rows. Ventrals 125–174; subcaudals single anteriorly or paired throughout, 21–68.

FIGURE 78.—Palestine Viper, *Vipera xanthina palaestinae.* This subspecies of the Near East viper is a leading cause of snakebite in its range. Photo by Erich Sochurek.

REFERENCES

ANDERSON, Steven C. 1963. Amphibians and Reptiles from Iran. Proc. Calif. Acad. Sci., vol. 31, pp. 417–498, figs. 1–15.

KHALAF, Kamal T. 1959. Reptiles of Iraq with Some Notes on the Amphibians. Ministry of Education: Baghdad, 96 pp., 40 figs.

MENDELSSOHN, H. 1963. On the Biology of the Venomous Snakes of Israel. Part I. Israel Jour. Zool., vol. 12, pp. 143–170, figs. 1–9. 1965. Part II, *ibid*, vol. 14, pp. 185–212.

MERTENS, Robert. 1952. Amphibien und Reptilien aus der Tuerkei. Rev. Fac. Sci. Univ. Istanbul, ser. B, no. 1, pp. 41–75.

MERTENS, Robert. 1965. Wenig bekannte "Seitenwinder" unter den Wüstenottern Asiens. Nat. u. Mus. vol. 95, pp. 346–352, figs. 1–5.

NIKOLSKY, A. M. 1916. Faune de la Russie. Reptiles. Petrograd. vol. 2, 349 pp., 8 pls.

SHAW, C. J. 1925. Notes on the Effect of the Bite of Mc Mahon's Viper (*E. macmahonii*). J. Bombay Nat. Hist. Soc., vol. 30: 485–486.

NOTES

Section 8

SOUTHEAST ASIA

Definition of the Region:

Includes the Indian Subcontinent; Tibet and other Himalayan States; the Chinese provinces of Sinkiang, Tsinghai, Szechwan, Yunnan, Kweichow, Kwangsi, Kwangtung, Hunan, Kiangsi, Fukien, Chekiang, and Hainan Island; Burma; Thailand; the Malay Peninsula; Laos; Cambodia; Viet Nam; Ceylon; Andaman and Nicobar Islands; Sumatra, Java and the Lesser Sunda Islands; Borneo and Celebes.

TABLE OF CONTENTS

TABLE 12.—DISTRIBUTION OF POISONOUS SNAKES IN SOUTHEAST ASIA

	W. Pakistan	India	Ceylon	E. Pakistan	Nepal, Sikkim, Bhutan	Kashmir, Tibet, Sinkiang, Tsinghai	Szechwan, Kweichow, Yunnan	S. E. China *	Burma	Thailand	Cambodia	Laos	Viet Nam	Malaya	Sumatra	Java	Lesser Sunda Is.	Andaman & Nicobar Is.	Borneo	Celebes	Djampea Is.
ELAPIDAE																					
Bungarus bungaroides		X																			
B. caeruleus	X	X	X																		
B. candidus										S	X	X	X	X	X	X				X	
B. ceylonicus			X																		
B. fasciatus		X		X				X	X	X	X	X	S	X	X	X			X		
B. flaviceps									X	S				X	X	X			X		
B. javanicus																X					
B. lividus		E		X	X																
B. magnimaculatus									X												
B. multicinctus							S	X	X				N								
B. niger		NE			X																
B. walli		E																			
Calliophis beddomei		S																			
C. bibroni		S	X																		
C. gracilis										S				X	X						
C. kelloggii							X	X				X	N								
C. macclellandii		NE			X		X	X	N	N	X	X	X								
C. maculiceps										X	X		S	N	X						
C. melanurus		X	X																		
C. nigrescens		X																			
Maticora bivirgata										S				X	X	X			X		
M. intestinalis										S				X	X	X			X	X	
Naja naja	X	X	X	X	?			X	X	X	X	X	X	X	X	X	X	X	X	X	
Ophiophagus hannah		X						X	X	X	X	X	X	X	X	X		X	X	X	
VIPERIDAE																					
Azemiops feae		?				SE	X		N				N								
Echis carinatus	X	X	N																		
Eristicophis macmahonii	NW																				
Pseudocerastes persicus	X																				
Vipera lebetina	N					SW															
V. russelii	E	X	X	X	SE	SW	S	X	X	N	?	?	N			X	X				
CROTALIDAE																					
Agkistrodon acutus								X													
A. halys	N	NW				N	N	?													
A. himalayanus	N	NW			X	X															

(Continued on page 118)

INTRODUCTION

In number and variety of species the snake fauna of southeast Asia is undoubtedly the richest in the world. It is the only region where virtually all major groups of snakes are represented. The richness of the fauna reflects partly the great variety of serpent habitats which range from semiarid slopes to fresh and salt water marshes, from alpine meadows to tropical rain forest. The region has been a major center of serpent evolution as well as one where some primitive types have survived.

Southeast Asia has also maintained for centuries a dense human population organized into a

MAP 9.—Section 8, Southeast Asia

succession of complex cultures and subcultures. Over this time span many kinds of snakes, including several venomous species, have developed a pattern of coexistence with man. Nowhere else do dangerous snakes and humans live in closer proximity in such numbers. This is the chief reason why southeast Asia has the world's highest

incidence of snakebites and snakebite deaths. This does not mean that snakes are everywhere apparent. Americans and Europeans visiting tropical Asia or living in its cities may never see a snake other than those exhibited by snake charmers. Extreme secretiveness is part of the snake's scheme for survival. But sometimes man and poisonous snake confront each other suddenly and unexpectedly to the everlasting disadvantage of one or both parties.

Vipers, cobras and their elapid allies, and sea snakes are all well represented in southeast Asia and all contribute to the snakebite problem. While few careful studies have been made, there is evidence that vipers, including pit vipers, are responsible for the greater number of snakebite cases while elapids are credited with fewer bites but a higher percentage of deaths. Sea snake bites are not uncommon in coastal villages.

GENERIC AND SPECIES DESCRIPTIONS

THE KRAITS

(*Bungarus*)

The name krait (of Hindhi origin) has been associated by English speaking peoples with a small venomous Indian snake. Actually there are several species of kraits and none of them are small; in fact, two reach lengths of about 7 feet. Kraits resemble many nonpoisonous snakes in general appearance. They have short rather flat heads only slightly wider than the neck. The eyes are small and dark, the pupils almost invisible in life. They are smooth scaled and glossy; most have a vivid pattern of crossbands. Three features of scutellation help distinguish kraits from other Asian snakes—a combination of all is diagnostic:

1. The vertebral row of scales is strongly enlarged, except in one rare species.

2. At least some of the subcaudals are undivided; in most species all are undivided.

3. The loreal shield is absent.

Kraits are strongly nocturnal, and their alert disposition by night differs from their quiet, almost stupid behavior by day. They cause few snakebites but the case fatality rate is very high.

Kraits lay eggs that are attended by the female. Their food consists largely of other snakes.

TABLE 12.—DISTRIBUTION OF POISONOUS SNAKES IN SOUTHEAST ASIA (continued)

	W. Pakistan	India	Ceylon	E. Pakistan	Nepal, Sikkim, Bhutan	Kashmir, Tibet, Sinkiang, Tsinghai	Szechwan, Kweichow, Yunnan	S. E. China *	Burma	Thailand	Cambodia	Laos	Viet Nam	Malaya	Sumatra	Java	Lesser Sunda Is.	Andaman & Nicobar Is.	Borneo	Celebes	Diampea Is.
CROTALIDAE (continued)																					
Agkistrodon hypnale		S	X																		
A. monticola							X														
A. nepa			X																		
A. rhodostoma										X	X	X	X	X	X	X					
A. strauchi						E	N														
Trimeresurus albolabris		NE			X			X	X	X	X	N	X	X	X	X					
T. cantori				X														X			
T. chaseni																			X		
T. convictus														X							
T. cornutus													X								
T. erythrurus		NE		X	X				X												
T. fasciatus				X																	
T. gramineus		X						X	X	X				X	X	X	X	X			X
T. hageni															X						
T. huttoni		S																			
T. jerdonii		NE		N		SE	X														
T. kanburiensis										X											
T. kaulbacki									N												
T. labialis																		X			
T. macrolepis		S																			
T. malabaricus		S																			
T. monticola		NE			X		S	X													
T. mucrosquamatus		NE					S	X	N	X	?	N	N								
T. popeorum		NE							X	X		X		X							
T. puniceus														X	X				X		
T. purpureomaculatus				S					X	SE	?			X	X	X	X	X	X		
T. stejnegeri		NE			X		S	X					N								
T. strigatus		S																			
T. sumatranus										S				X	X		X		X	X	
T. tonkinensis								?					N								
T. trigonocephalus			X																		
T. wagleri										S				X	X	X	X		X	X	

Certain groups of adjoining nations or provinces are here treated as units. The symbol X indicates distribution of the species is widespread within the unit. Restriction of a species to part of a unit is indicated appropriately (SW = Southwest, etc.). The symbol ? indicates suspected occurrence of species within the area without valid literature.

*Kwangsi, Kwangtung, Hunan, Kiangsi, Fukien, Chekiang provinces and the island of Hainan.

KEY TO GENERA

1. A. Loreal pit present_____ 2
   B. Loreal pit absent_____ 3
2. A. No enlarged crown shields_____ *Trimeresurus*
   B. Five to nine enlarged crown shields_____ *Agkistrodon*
3. A. Crown of head with large shields_____ 4
   B. Crown of head with small scales_____ 12
4. A. Tail paddle-shaped_____ Sea snake
                                                    (See Chapter VIII)
   B. Tail not paddle-shaped_____ 5
5. A. Loreal scale present_____ 6
   B. Loreal scale absent_____ 7
6. A. Movable fangs present_____ *Azemiops*
   B. No fangs present_____ NP*
7. A. Vertebral scale row enlarged and at least some
        subcaudals undivided_____ *Bungarus*
   B. Without the above combination of characters_____ 8
8. A. Occipital shields present; anterior subcaudals
        undivided_____ *Ophiophagus*
   B. Not as above_____ 9
9. A. Dorsal scales smooth_____ 10
   B. Dorsal scales keeled_____ NP
10. A. Hood seen in life; body scales in 19 or more rows
         anteriorly_____ *Naja*
    B. No hood; slender snakes with no more than 15
         scale rows anteriorly_____ 11
11. A. Venom glands in normal position; anal shield
         usually divided_____ *Calliophis*
    B. Elongate venom glands extending well into
         body; anal shield entire_____ *Maticora*
12. A. Ventrals extending full width of belly_____ 13
    B. Ventrals not extending full width of belly or
         absent_____ NP
13. A. Subcaudals undivided_____ *Echis*
    B. Subcaudals divided_____ 14
14. A. Hornlike process above eye_____ *Pseudocerastes*
    B. No hornlike process above eye_____ 15
15. A. Rostral flanked by enlarged scales; ventrals with
         lateral ridge_____ *Eristicophis*
    B. Rostral not flanked by enlarged scales; ventrals
         without lateral ridge_____ *Vipera*

* NP = Nonpoisonous

ELAPIDAE: Genus *Bungarus* Daudin, 1803.

Kraits.

Twelve species are recognized; all inhabit the region of southeast Asia. Occasional individuals of *B. fasciatus* attain lengths of 7 feet. Most species are of moderate (4 to 5 feet) length, but all are considered extremely dangerous.

Definition: Head small, flattened, slightly distinct

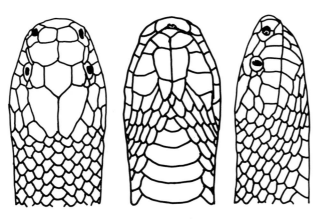

FIGURE 79.—Head Scales of Krait (*Bungarus*). Note the small eye and the nasal in broad contact with the single preocular. Redrawn from Maki, 1931.

from neck; no distinct canthus. Body moderately slender, cylindrical; tail short.

Eyes small; pupils round or vertically subelliptical.

Head scales: The usual 9 on the crown; frontal broad. Laterally, nasal in broad contact with single preocular.

Body scales: Dorsals smooth, vertebral row enlarged and hexagonal (strongly so except in *B. lividus*), in 13–17 oblique rows at midbody. Ventrals 193–237; anal plate entire; subcaudals single or paired (all paired only in some specimens of *B. bungaroides*), 23–56.

Maxillary teeth: Two large tubular fangs with external grooves followed, after an interspace, by 1–4 small, feebly-grooved teeth.

Indian Krait, *Bungarus caeruleus* (Schneider).

Identification: Body cylindrical with slight even taper; tail with pointed tip. All subcaudals undivided.

Jet black to dark brown with a series of narrow white or yellow crossbands that tend to be in pairs and often fade out or break up on the anterior quarter of the body; upper lip white or yellow; belly an immaculate white (see plate VI, fig. 5).

Average adult length 3 to 4 feet; maximum slightly over 5 feet.

Distribution: Essentially restricted to India and parts of West Pakistan. Found in a variety of habitats at low and moderate elevations preferring rather dry open country. Often found near human habitations and

frequently enters poorly constructed or delapidated buildings.

Remarks: Indian kraits usually prowl on hot humid nights and are quite agile in their movements. When alarmed they coil loosely with the body slightly flattened and head concealed. They make jerky movements and may elevate the tail. They do not strike but often make a quick snapping bite. During the day they are much more lethargic.

This is the most dangerous of the kraits for it has a venom of very high toxicity for man—the lethal dose is estimated at about 4 mg. Bites are rare but the fatality rate in one series of 35 cases was 77 percent. Antivenin is produced by the Central Research Institute, Kasauli, India; and the Haffkine Institute, Bombay, India.

Ceylon Krait, *B. ceylonicus* (Günther).
and
Malayan Krait, *B. candidus* (Linnaeus).

Description: These two kraits are very similar to the Indian krait in general appearance, but have fewer crossbands (15–25 versus 35–55 for *caeruleus*). The bands are wide in *candidus*, narrow and often broken in *ceylonicus*.

Many-banded Krait, *B. multicinctus* Blyth.

Identification: Very similar to the Indian krait but the light crossbands are not in pairs and the underside may show dark mottling. It is a little smaller than *caeruleus* having an average length of 35 to 45 inches and maximum of less than 5 feet.

FIGURE 80.—Many-banded Krait, *Bungarus multicinctus.* From a painting. (See also plate V, figure 2).

Distribution: Burma through southern China to Hainan and Taiwan. It frequents wooded or grassy places near water and may be found in villages and suburban areas. It is common in rice paddies.

Remarks: Active on damp or rainy nights; inoffensive in disposition as a rule. Toxicity of the venom for animals is extremely high (LD$_{50}$ about 0.1 mg. per kilo). Bites by this krait are seen regularly in Taiwan, but the case fatality rates are less than half those reported for India. Antivenin is produced by the Taiwan Serum Vaccine Laboratory, Taipei.

Banded Krait, *Bungarus fasciatus* (Schneider).

Identification: A marked vertebral ridge giving a permanently emaciated appearance, and a distinctly blunt tail are characteristic of this species.

Pattern of alternating light and dark bands encircling the body and of almost equal width. The light bands are usually bright yellow, occasionally white, pale brown or orange; the dark bands are black.

Average length 4 to 5 feet; maximum about 7 feet.

Distribution: Eastern India to southern China and south through much of Malaysia and Indonesia. Occurs in rather open country to elevations of about 5,000 feet, often found near water.

Remarks: This is such a surprisingly quiet, inoffensive snake that it is believed harmless over much of the territory where it is found. When annoyed it curls up,

FIGURE 81.—Banded Krait, *Bungarus fasciatus*. The blunt tail is typical. Photo by New York Zoological Society.

hides its head beneath its coils, and makes jerky flinching movements but does not bite except in rare instances. Cases of snakebite due to the banded krait are almost unknown. Its venom is of lower toxicity for animals than that of some other kraits. Antivenin is produced by the Institut Pasteur, Paris; the Institut Pasteur Bandung, Indonesia and the Queen Saovabha Memorial Institute, Bangkok.

Red-headed Krait, *Bungarus flaviceps* Reinhardt.

Identification: General appearance like the banded krait, but tail only slightly blunt; anterior subcaudals entire, posterior ones divided.

Very striking and distinctive coloring—head and tail bright red, body black with narrow bluish white stripe low on side, and sometimes a narrow orange stripe or row of dots down middle of back.

Size about the same as the banded krait.

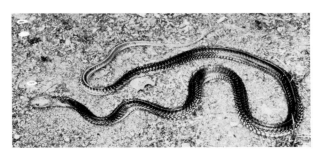

FIGURE 82.—Red-headed Krait, *Bungarus flaviceps*. Both the head and the tail are bright red in this species. Photo by D. Dwight Davis.

Distribution: Southern Burma to Viet Nam and south through Malaysia and larger islands of Indonesia. Inhabits jungle mostly in hilly or mountainous country. A rare snake.

Remarks: Apparently much like the banded krait in behavior. No study of its venom has been done nor are there records of its biting man. Antivenin is produced by Institut Pasteur, Paris.

ELAPIDAE: Genus *Calliophis* Gray, 1834.

Oriental coral snakes.

Thirteen species are recognized; all inhabit the region of southeastern Asia. Most are small species but a few exceed 3 feet in length. At least the larger individuals are considered dangerous.

Definition: Head small, not distinct from body. Body cylindrical, slender and elongated; tail short.

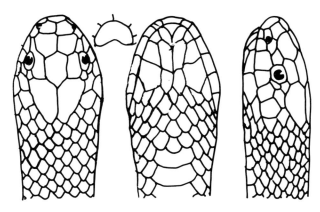

FIGURE 83.—Head Scales of Oriental Coral Snake, *Calliophis*. Redrawn from Maki, 1931.

Eyes small to moderate in size; pupils round.

Head scales: The normal 9 on the crown; rostral broad and rounded, no canthus. Laterally, nasal in contact with single preocular or separated from it by prefrontal; preocular absent in *C. bibroni*.

Body scales: Dorsals smooth, in 13–17 nonoblique rows throughout body. Ventrals 190–320; anal plate entire or divided; subcaudals usually paired, occasionally single in *C. macclellandii*, 12–44.

Maxillary teeth: Two large tubular fangs with external grooves followed, after an interspace, by 0–3 small teeth.

Oriental Coral Snakes, *Calliophis*.

Identification: This genus of generally small snakes includes the species formerly in the genus *Hemibungarus*. They all have a small head which is barely or not distinct from the neck; a long body which is slim and cylindrical with little taper; 13 or 15 scale rows at midbody, rarely 17; a short tail; and a smooth polished appearance. Like the American coral snakes, they are difficult to distinguish from some nonpoisonous snakes. Absence of the loreal shield in the coral snakes helps, but similar nonpoisonous snakes may lack this shield. Divided subcaudals and the absence of enlarged vertebral scales distinguish them from kraits. Cobras are larger and have a quite different body build.

Color and pattern show marked individual and species variation. Representative patterns are exemplified by *Calliophis macclellandii* which is russet to pink with narrow, widely separated black crossbands and a wide cream band across the base of the head, and by *C. sauteri* which is brown to crimson with 3 longitudinal black stripes and a narrow cream headband. (See plate V, fig. 6; plate VI, fig. 1.)

Distribution: The genus occurs throughout southeast Asia including the Philippines, Malaysia, and Taiwan. They are snakes of forest country ranging well into the mountains but they avoid dry terrain. Occasionally they have been collected in suburban areas.

Remarks: The Oriental coral snakes are generally considered rare, but this may only reflect their very secretive nature. They have been found under logs or ground litter and occasionally in the open at night.

They are quiet reptiles apparently very reluctant to bite. Some authorities consider them essentially harmless, but it may be recalled that the North American coral snake once had this reputation.

ELAPIDAE: Genus *Maticora* Gray, 1834.

Long-glanded coral snakes.

Two species are found in the region of southeastern Asia: from Thailand and the Philippines to Sumatra, Java, Borneo, and Celebes. These snakes are relatively small and slender but individuals of one species (*M. bivirgata*) occasionally approach 5 feet in length; such individuals are believed to be capable of inflicting a dangerous bite.

Definition: Head small and not distinct from body. Body cylindrical, slender and elongated; tail short.

Eyes small to moderate; pupils round.

Head scales: The usual 9 on the crown; no canthus; rostral broad and rounded. Laterally, nasal in broad contact with single preocular; eye in contact with supralabial row.

Body scales: Dorsals smooth, in 13 nonoblique rows throughout body. Ventrals 197–293; anal plate entire; subcaudals paired, 15–50.

Maxillary teeth: Two large tubular fangs; no other teeth on the bone.

Remarks: The only consistent difference between these snakes and those of the genus *Calliophis* is that *Maticora* has elongated venom glands that extend posteriorly for about one-third of the body length. The heart has been pushed back to the middle third of the body, where it can be felt (in preserved specimens) as a hard object, thus identifying the genus.

Color and pattern exhibit much variation. In general they are dark brown to blue black above with narrow light stripes of yellow, red, pale blue, violet, or white. The belly of the common long-glanded snake (*Maticora intestinalis*) is black and white; the tail red barred with black. In the red-bellied long-glanded snake (*M. bivirgata*) the entire head, tail and belly are bright red.

Figure 84.—Long-glanded Snake, *Maticora bivirgata*. Photo by D. Dwight Davis

These are secretive, inoffensive snakes. When disturbed they squirm about violently often curling and elevating the tail to display their bright color. The behavior is characteristic of some Asian and American coral snakes and may occasionally be seen in kraits. It is likewise demonstrated by several unrelated kinds of nonpoisonous snakes.

Although they rarely bite, the long-glanded snakes must be considered potentially dangerous. Serious poisoning has resulted from the bite of *M. intestinalis* and death from the bite of *M. bivirgata*. No antivenins are produced against venoms of long-glanded snakes or Oriental coral snakes.

COBRAS

The cobras are at once the best and most poorly known of Asia's poisonous snakes. Except for the very distinct king cobra, all central and south Asian populations are regarded as subspecies of *Naja naja*. There are, however, some significant differences in fang type, pattern, color, behavior, and venom composition among these forms, hence they will be considered separately.

The Asian cobras are at home in many types of terrain, only desert and dense rain forest being generally avoided. Flat country with high grass and scattered groves of trees is an optimum habitat. Rice fields and other sorts of agricultural land may support many cobras, and they are often common around villages and cities. Here they may be found in crumbling walls, old buildings, and gardens.

In western India and Pakistan cobras are more active by day—usually in the evening and early morning—while in the countries to the east they show a greater tendency to be nocturnal. They are timid when encountered in the open and seek to escape. When cornered they rear up and spread their hoods, but biting seems to be almost a last resort. The snakes frequently strike with the mouth closed. They are most dangerous when surprised at close quarters. In biting, they tend to hold on, chewing savagely. Although the fangs of Malayan, Indonesian, and Philippine cobras are modified for spraying venom at the eyes, this behavior seems to be uncommon, at least when the snake is confronted by a human foe.

The hood identifies a living cobra. Although some nonpoisonous Asian snakes flatten the neck slightly when alarmed, none do so to such a marked degree as do the Asian *Naja*. The hood of the king cobra is much narrower. Identification of dead cobras is more difficult. Among the more useful scale characteristics are absence of the loreal shield and the presence of a large third supralabial which touches both the eye and the nasal shield. This combination is seen elsewhere only in some of the Oriental coral snakes. Color and pattern are extremely variable and will be discussed in the following paragraphs; however, most Asian *Naja* have conspicious dark bars or spots on the underside of the neck at about the level of the hood. This is not seen in many nonpoisonous snakes that might be confused with cobras.

Large cobras may have a great quantity of venom—sometimes 500 to 600 mg.—and the lethal dose for man is estimated at not more than 20 mg. In spite of this, many persons recover from bites without effective treatment. Evidently the snakes may inject little venom when biting defensively. Some cobra bites are accompanied by extensive necrosis with little systemic effect. Such bites have been reported in Malaya and elsewhere in southeast Asia and also in west Africa. The strongly lethal component of cobra venom can be separated from that component producing local necrosis, and it appears that venom of some pop-

FIGURE 85.—Indian Cobra, *Naja naja naja*. The hood pattern of two spots is distinctive. Photo by Erich Sochurek.

ulations or individual snakes is high in necrotizing factor but low in neurotoxin.

Asian cobras feed upon almost any kind of vertebrate small enough to be swallowed. Their fondness for rats helps explain their abundance near human habitations. Cobras lay eggs, 10 to 20 being an average clutch. The female and occasionally the male remain with the eggs and may defend them.

Antivenin sources: Antivenin against venom of Asian cobras is produced by the Behringwerke, Marburg-Lahn, Germany; Central Research Institute, Kasauli, India; Haffkine Institute, Bombay, India; Tashkent Institute, Moscow (*Naja n. oxiana*); Queen Saovabha Memorial Institute, Bangkok, Thailand; Commonwealth Serum Laboratories, Victoria, Australia (Malay *Naja*); Taiwan Serum Vaccine Laboratory, Taipei (*Naja n. atra*); State Razi Institute, Tehran, Iran (*Naja n. oxiana*); Department of Health, Manila, Philippines (*Naja n. philippinensis*); Institut Pasteur, Bandung, Indonesia (*Naja n. sputatrix*).

ELAPIDAE: Genus *Naja* Laurenti, 1768.

Cobras.

Six species are recognized; all are African except the Asiatic cobra, *Naja naja*, and range throughout the African continent except for the drifting sand areas of the Sahara region. They are snakes of moderate (4 feet) to large (8 feet) size, with large fangs and toxic venom. The species, *N. nigricollis*, "spits" its venom at the eyes of an aggressor; it is found in the southern part of the region of north Africa. The Egyptian cobra (*Naja haje*) and the western subspecies of the Asiatic cobra (*Naja naja oxiana*) are found in the Near and Middle East region. The Asiatic cobra, *N. naja*, is the only species in this region.

Definition: Head rather broad, flattened, only slightly distinct from neck; snout rounded, a distinct canthus. Body moderately slender, slightly depressed, tapered; neck capable of expansion into hood; tail of moderate length.

Eyes moderate in size; pupils round.

Head scales: The usual 9 on the crown; frontal short; rostral rounded. Laterally, nasal in contact with the one or two preoculars.

Body scales: Dorsals smooth, in 17–25 oblique rows at midbody, usually more on the neck, fewer posteriorly. Ventrals 159–232; anal plate entire; subcaudals 42–88, mostly paired.

Maxillary teeth: Two rather large tubular fangs with external grooves followed, after an interspace, by 0–3 small teeth.

Indian Cobra, *Naja naja naja* (Linnaeus).

Identification: Adults brown or black, uniform or with variegation produced by rows of dappled or bicolored-scales. There is a "spectacle" type hood mark present, except in black individuals where it is apparently obscured by pigment. Belly is light anteriorly becoming clouded posteriorly, or generally dark with light areas on neck. Young paler and more variegated. In populations where the adults are uniform black, the young show a hood mark.

Average length 4 to 5 feet, maximum about 6½ feet. Sexes of about equal size.

Distribution: Most of the Indian Subcontinent exclusive of the extreme northwest and region east of the Ganges delta; Ceylon.

Oxus Cobra, *Naja naja oxiana* (Eichwald).

Identification: Adults brown, sometimes with traces of wide dark crossbands; hood mark never present; belly pale with dark bars on neck. Young tan or buff with regular dark crossbands; no hood mark. The hood in this form is noticeably narrower than in other Asian *Naja*.

Length about the same as the Indian cobra.

FIGURE 86. Oxus Cobra, *Naja naja oxiana*. This western representative of the Asiatic Cobra occurs in northeastern Iran and in Afghanistan. Photo by Allan Roberts.

Distribution: Northern frontier of West Pakistan across Afghanistan and into eastern Iran and southern parts of Russian Asia. Avoids desert areas; occurs in mountains to about 7,500 feet.

Monocellate Cobra, *Naja naja kaouthia* Lesson.

Identification: Brown or black usually speckled or variegated with white or pale yellow and often showing alternate wide and narrow transverse dark bands; dorsal hood mark a pale circle edged with black and enclosing 1 to 3 dark spots; ventral hood mark a pair of dark spots or a wide dark band. Young darker than adults and with more vivid crossbands.

Distribution: West Bengal, East Pakistan, Assam and Burma; Thailand; Malaya and southwest China, mostly in lowlands.

Chinese Cobra, *Naja naja atra* Cantor.

Identification: Adults grayish brown, olive or blackish with widely spaced narrow light bands sometimes in pairs; hood marks variable but usually similar to the monocellate cobra; belly pale sometimes with brown mottling. Young black with distinct whitish crossbands. Slightly smaller than the Indian cobra; maximum length about 5½ feet. (See plate VI, fig. 2).

Distribution: Thailand and south China east to Viet Nam, Hainan and Taiwan.

Malay Cobra, *Naja naja sputatrix* Boie.

Identification: Brown, gray or black without definite pattern on body; hood marks as in the monocellate cobra or dorsal mark absent; belly dark sometimes with white blotches on the throat. In this race of the Asian cobra, the discharge orfice of the fang is small and well short of the tip. This type of fang is associated with the habit of spraying or "spitting" venom, and such behavior has been reported for the Malayan cobra.

Average length 40 to 50 inches; maximum about 60 inches.

Distribution: The Malay peninsula and most of the larger islands of Indonesia.

Borneo Cobra, *Naja naja miolepis* Boulenger.

Identification: Black or very dark brown above without a dorsal hood marking; belly yellow to dark gray. Young with widely spaced white or yellow crossbands and a chevron-shaped light mark behind the head. The maximum length is about 55 inches.

Distribution: Borneo, Palawan (Philippines). *Naja naja samarensis* of the Visayan Islands of the Philippines is very similar.

Philippine Cobra, *Naja naja philippinensis* Taylor.

Identification: Light brown or olive above without hood marking; cream to light brown below. Young darker with reticulate pattern of light lines. Size about the same as the Borneo cobra.

Distribution: Luzon and Mindoro, Philippines.

ELAPIDAE: Genus *Ophiophagus* Günther, 1864.

King cobra.

A single species, *O. hannah*, is found in southeastern Asia and the Philippines. It attains a length of 16 to 18 feet, and is considered one of the world's most dangerous snakes.

Definition: Head relatively short, flattened, moderately distinct from neck; snout broad, rounded, canthus indistinct. Body slender, tapering, neck region capable of expanding into small hood; tail long.

Eyes moderate in size; pupils round.

Head scales: The usual 9 on the crown, plus a pair of large occipitals in contact with one another behind the parietals. Laterally, nasal in narrow contact with elongate preocular.

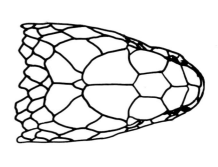

FIGURE 87.—Head Scales of King Cobra, *Ophiophagus hannah*. The large scutes (occipitals) behind the parietal scutes identify this genus. Redrawn from Maki, 1931.

Body scales: Dorsals smooth, in 15 oblique rows at midbody and posteriorly, more (17–19) on neck. Ventrals 240–254; subcaudals 84–104, the anterior ones single, the remainder paired.

Maxillary teeth: Two relatively short fangs (about ½ inch in a large specimen) with external grooves followed, after an interspace, by 3 small teeth.

King Cobra, Hamadryad, *Ophiophagus hannah* (Cantor).

Identification: The great size is an important recognition feature. Adults in most parts of the range measure 7 to 13 feet and are larger than any Asian snakes except the pythons and exceptional specimens of the nonpoisonous keeled rat snake (*Zaocys*) which may reach 12 feet. Smaller king cobras may be recognized by the presence of large occipital shields, a unique feature of the species. The hood is proportionally narrower than in Asian *Naja*.

Adults olive, brown or greenish yellow becoming darker on the tail; head scales edged with black; throat yellow or orange sometimes with black markings. Young black with buff, white or yellow chevron-shaped narrow crossbands. Adult snakes from East Bengal, Burma and Thailand retain the crossbands especially on the posterior half of the body.

Distribution: Peninsular India to the Himalayan foothills thence eastward across southeastern China and regions to the south; the Philippines and larger islands of Indonesia. In the western and northern part of its range largely confined to hilly jungle to elevations of 6,000 feet. In Malaya and Thailand found in fairly open country and in cultivated areas. Nowhere is it very plentiful.

Remarks: King cobras are active diurnal snakes. They are primarily terrestrial, but are sometimes found in trees and in the water. While they have been reported to make unprovoked attacks, such behavior is extremely unusual. If cornered or injured they can be very dangerous, but they frequently give little evidence of hostility when encountered. When angry they give a deep resonant hiss similar to the growl of a small dog.

FIGURE 88.—King Cobra, *Ophiophagus hannah.* The hood is much smaller than in the Asiatic Cobra, *Naja naja.* Photo by San Diego Zoo.

The king cobra is unique among snakes in constructing an elaborate nest of dead leaves and other decaying vegetation. There are two chambers, one for the eggs, the other occupied by the female snake. The male may also remain nearby. Nesting cobras frequently but not invariably defend their eggs.

King cobra bites in man appear to be most infrequent; indeed there seems to be no adequate account of the symptoms of envenomation. Venom of the king cobra shows marked antigenic differences from *Naja* venom and is not well neutralized by *Naja* antivenin. Its toxicity for mammals is less than that of Asiatic *Naja* venoms. King cobra antivenin is produced by Queen Saovabha Memorial Institute, Bangkok.

VIPERIDAE: Genus *Azemiops* Boulenger, 1888.

Fea's viper.

A single species, *A. feae* Boulenger, is known from the mountains of southeastern Asia. It is a small species, less than 3 feet in length, and its danger to man is unknown.

Definition: Head somewhat flattened, distinct from neck; snout broad and short, canthus obtuse. Body cylindrical, moderately slender; tail short.

Eyes moderate in size; pupils vertically elliptical.

Head scales: The usual 9 scutes on the crown; rostral broad, frontal broad. Laterally, eye in contact with supralabial row; nasal separated from preoculars by small squarish loreal.

Body scales: Dorsals smooth, in 17 nonoblique rows at midbody, fewer (15) posteriorly. Ventrals rounded, 180–189; subcaudals paired throughout or a few anterior ones single, 42–53.

VIPERIDAE: Genus *Echis* Merrem, 1820.

Saw-scaled vipers.

Two species are recognized. One (*E. coloratus*) is restricted to eastern Egypt and the Arabian peninsula. The other (*E. carinatus*) ranges from Ceylon and southern India across western Asia and north Africa southward into tropical Africa. Although neither attains a length of 3 feet, they possess a highly toxic venom and are responsible for many deaths. When disturbed they characteristically inflate the body and produce a hissing sound by rubbing the saw-eged lateral scales against one another. (see p. 83, fig. 52). This same pattern of behavior is shown by the nonpoisonous egg-eating snakes *Dasypeltis.*

Definition: Head broad, very distinct from narrow neck; canthus indistinct. Body cylindrical, moderately slender; tail short.

Eyes moderate in size; pupils vertically elliptical.

Head scales: A narrow supraocular sometimes present; otherwise crown covered with small scales, which may be smooth or keeled. Rostral and nasals distinct. Laterally eye separated from labials by 1–4 rows of small scales; nasal in contact with rostral or separated from it by a row of small scales.

Body scales: Dorsals keeled, with apical pits, lateral scales smaller, with serrate keels, in 27–37 oblique rows at midbody. Ventrals rounded, 132–205; subcaudals single, 21–52.

VIPERIDAE: Genus *Eristicophis* Alcock and Finn, 1897.

A single species, *E. macmahonii* Alcock and Finn, is known from the desert areas of southeastern Iran, Afghanistan, and West Pakistan. It is a rather small snake, less than 3 feet in length. However, fatal cases attributed to this species (Shaw, 1925) and a recent serious bite indicate that it is a dangerous snake with venom similar to that of *Echis* (see p. 110, fig. 75).

Definition: Head broad and flattened, very distinct from neck; snout broad and short, canthus not distinct. Body slightly depressed, moderately to markedly stout; tail short.

Eyes moderate in size; pupils vertically elliptical.

Head scales: Crown covered by small scales; rostral broad, bordered dorsally and laterally by greatly enlarged nasorostral scales. Laterally, eye separated from labials by 3–4 rows of small scales; nasal separated from rostral by nasorostral scale.

Body scales: Dorsals keeled, short, in 23–26 vertical rows at midbody. Ventrals with lateral keels, 140–148; subcaudals paired, without keels, 29–36.

VIPERIDAE: Genus *Pseudocerastes* Boulenger, 1896.

False-horned viper.

A single species is recognized (see Remarks, pp. 110–111). It ranges from Sinai and the Arabian Peninsula eastward to West Pakistan. It attains a length of 3 feet and is considered dangerous.

Definition: Head broad, very distinct from neck; snout short and broadly rounded; nostrils dorsolateral, valves present.

Eyes small to moderate; pupils vertically elliptical.

Head scales: Crown covered with small imbricate scales; an erect hornlike projection covered with imbricate scales above eye. Laterally, nasals separated from rostral by small scales; eye separated from labials by 3–4 rows of small scales.

Body scales: Dorsals weakly to moderately keeled, in 21–25 nonoblique rows at midbody. Ventrals 134–158; subcaudals paired, 35–48.

VIPERIDAE: Genus *Vipera* Laurenti, 1768.

True adders.

Eleven species are recognized. This is an especially variable group, with some members that are small and relatively innocuous (e.g., *V. berus*) and others that are extremely dangerous (*V. lebetina*, *V. russelii*). They are found from northern Eurasia throughout that continent and into north Africa. One species ranges into the East Indies (*V. russelii*), and two are found in east Africa (see Remarks under *V. superciliaris*). Russell's viper and the Levantine viper (p. 111) are the only members of this genus in the region.

Definition: Head broad, distinct from narrow neck; canthus distinct. Body cylindrical, varying from moderately slender to stout; tail short.

Eyes moderate in size to small; pupils vertically elliptical.

Head scales: Variable: one species (*V. ursinii*) has all 9 crown scutes, most species have at least the supraoculars, but even these are absent in one (*V. lebetina*); head otherwise covered with small scales. Laterally, nasal in contact with rostral or separated by a single enlarged scale (the nasorostral), eye separated from supralabials by 1–4 rows of small scales.

Body scales: Dorsals keeled, with apical pits, in 19–31 nonoblique rows at midbody. Ventrals rounded, 120–180; subcaudals paired, 20–64.

Russell's Viper, *Vipera russelii* (Shaw).

Identification: Head wide, rather long; no enlarged plates on crown; no loreal pit; scales keeled. These features and the bold distinctive pattern readily distinguish this reptile from most other Asian snakes. It may be closely imitated by the harmless Russell's sand boa (*Eryx conicus*), however this species has narrow ventrals (less than the width of the belly) and a very short tail.

Color deep yellow, tan, or light brown with 3 rows of large oval dark black-ringed spots which may be narrowly edged with white; the spots of the middle row often fuse on the latter half of the body; light V or X shaped mark on top of the head; belly pinkish brown to white with black spots.

Populations of this viper from Indonesia, Taiwan, China and Thailand are more grayish or olive; there are small spots between the rows of large spots and the belly is suffused with gray posteriorly.

Average length 40 to 50 inches; maximum 65 inches; males larger than females. The island races average smaller.

FIGURE 89.—Russell's Viper, *Vipera russelii*. The oval black-bordered markings are typical. Photo by R. Van Nostrand. (See also plate VI, figure 3.)

Distribution: Eastern West Pakistan, most of India, Burma, and Ceylon; parts of Thailand, southeast China, Taiwan, and a few islands of Indonesia. Over most of its range, a snake of open grassy or brushy country often common around cultivated fields and villages. Occurs in lowlands, but avoids permanently marshy areas. Primarily a hill or mountain snake in some places and has been recorded at 7000 feet elevation.

Remarks: Mainly nocturnal but occasionally active by day in cool weather. Crawls slowly and is rather phlegmatic in disposition. Hisses loudly when disturbed and strikes with great force and speed.

Russell's viper is very prolific giving birth to 20 to 60 young. As is true of many snakes, the young are more irritable than the adults.

Russell's viper is a leading cause of snakebite accidents in India and Burma, but the case fatality rate is lower than in bites by kraits, cobras and saw-scaled vipers. The lethal dose of Russell's viper venom for man is estimated at 40–70 mg.; a large snake yields 150–250 mg. Antivenins are produced by the Behringwerke, Marburg-Lahn, Germany; Central Research Institute, Kasauli, India; Haffkine Institute, Bombay, India; Queen Saovabha Memorial Institute, Bangkok.

CROTALIDAE: Genus *Agkistrodon* Beauvois, 1799.

Moccasins and Asian pit vipers.

Twelve species are recognized. Three of these are in North and Central America; the others are in Asia, with one species. *A. halys* (Pallas) ranging westward to southeastern Europe. The American copperhead (*A. contortrix*) and the Eurasian mamushi and its relatives (*A. halys*) seldom inflict a serious bite but *A. acutus* and *A. rhodostoma* of southeastern Asia, as well as the cottonmouth (*A. piscivorus*) of the southeastern United States, are dangerous species (see p. 136 for description of *A. acutus*).

Definition: Head broad, flattened, very distinct from narrow neck; a sharply-distinguished canthus. Body cylindrical or depressed, tapered, moderately stout to stout; tail short to moderately long.

Eyes moderate in size; pupils vertically elliptical.

Head scales: The usual 9 on the crown in most species; internasals and prefrontals broken up into small scales in some Asian forms; a pointed nasal appendage in some. Laterally, loreal pit separated from labials or its anterior border formed by second supralabial. Loreal scale present or absent.

Body scales: Dorsals smooth (in *A. rhodostoma* only) or keeled, with apical pits, in 17–27 nonoblique rows. Ventrals 125–174; subcaudals single anteriorly or paired throughout, 21–68.

Malayan Pit Viper, *Agkistrodon rhodostoma* (Boie).

Identification: Head triangular, snout pointed, facial pit present. The only Asian pit viper with large scales on the crown and smooth body scales.

Middle of back reddish or purplish brown, sides paler with dark speckling; series of dark brown crossbands, narrow in midline, wider on sides, edged with white or buff; belly pinkish white mottled with brown; top of head dark brown, sides light pinkish brown, the colors separated by a white stripe that passes just above the eye.

Average length 23 to 32 inches; maximum about 40 inches.

Distribution: Thailand, northern Malaysia, Cambodia, Laos, Viet Nam, Java, Sumatra—apparently requires climate with well-marked wet and dry seasons. Frequents forests generally at low elevations; common on rubber plantations.

Remarks: A bad tempered snake, quick to strike if disturbed. In northern Malaysia it causes approxi-

FIGURE 90.—Malayan Pit Viper, *Agkistrodon rhodostoma*. This smooth-scaled pit viper is the source of many bites in southeast Asia. Photo by New York Zoological Society.

mately 700 snakebites annually with a death rate of about 2 percent. Weeders and tappers on rubber estates are most frequently bitten. The snake is remarkably sedentary and has often been found at the site of an accident after several hours.

This is another of the oviparous vipers. The eggs are guarded by the female. Antivenin is produced by the Institut Pasteur, Paris; the Institut Pasteur, Bandung, Indonesia; and Queen Saovabha Memorial Institute, Bangkok.

Hump-nosed Viper, *Agkistrodon hypnale* (Merrem).

Identification: Of typical viperine build with stout body and wide head with facial pit; snout pointed and turned up; large frontal and parietal shields but shields of snout small and irregular.

Grayish, heavily powdered or mottled with brown; double row of large dark spots; belly yellowish or brownish with dark mottling; tip of tail reddish or yellow.

Average length 12 to 18 inches.

FIGURE 91.—Hump-nosed Viper, *Agkistrodon hypnale*. Photo by Edward H. Taylor (Preserved specimen).

Distribution: Southern India and Ceylon. Inhabits dense jungle and coffee plantations in hilly country.

Remarks: Often found by day coiled in bushes. It is irritable and vibrates the tail when annoyed. Bites by this snake are seen fairly frequently, but serious poisoning has not been reported. There is no antivenin.

ASIAN LANCE-HEADED VIPERS
(*Trimeresurus*)

This large genus, containing some 30 species, is closely related to tropical American lanceheads (*Bothrops*). All have large triangular heads much wider than the neck. Presence of the facial pit and absence of large plates* on the top of the head (fig. 96) distinguish them from most other snakes within their range. The pupils of the eye are elliptical; the subcaudals may be divided or undivided.

Bites by these snakes are quite frequent; however, the fatality rate is very low. The American polyvalent Crotalid Antivenin (Wyeth, Inc., Philadelphia) shows neutralizing activity against venoms of several Asian lanceheads. It should be used if specific antivenin is not available.

CROTALIDAE: Genus *Trimeresurus* Lacépède, 1804.

Asian lance-headed vipers.

About 30 species are currently recognized. All are found in southeast Asia and the adjacent island chains. The large species are dangerous; many of the smaller kinds can deliver a venomous bite which is very painful, but seldom if ever fatal (see pp. 137–138 for description of other species that enter this region).

There are 3 general groups of these snakes:

1. Large, long-bodied and long-tailed terrestrial snakes that are often brightly-colored with contrasting patterns;

2. Small short-bodied and short-tailed terrestrial snakes, commonly with dull patterns of brown blotches;

3. Small, moderately long-bodied arboreal snakes with prehensile tail, body coloration tending toward unicolor greens, light browns, or light speckles.

Definition: Head broad, flattened, very distinct from narrow neck; canthus obtuse to sharp. Body cylindrical to moderately compressed, moderately slender to stout; tail short to moderately long.

Eyes small to moderate in size; pupils vertically elliptical.

Head scales: Supraoculars present, a pair of internasals often present; remainder of crown covered with small scales. Laterally, a nasal pore in prenasal, 2 enlarged preoculars, eye separated from supralabials by 1–4 rows of small scales.

---

* Present in *T. macrolepis* of south India.

Body scales: Dorsals feebly to strongly keeled, in 13–37 nonoblique rows. Ventrals 129–231; subcaudals paired, 21–92.

Chinese Green Tree Viper, *Trimeresurus stejnegeri* Schmidt.

Identification: One of a group of very similar arboreal pit vipers found throughout much of tropical Asia. All are slender to moderately stout snakes with prehensile tails. In this species the first upper lip shield is not fused with nasal shield and the dorsal scales are keeled.

Body bright green to chartreuse above, yellow to pale green ventrally; white or yellow line on side of body edged with reddish in male; upper lip yellow or green; iris of eye orange to coppery; end of tail reddish.

Average length about 20 inches.

FIGURE 92.—Chinese Green Tree Viper, *Trimeresurus stejnegeri*. Photo by Isabelle Hunt Conant. (See also plate IV, figure 2.)

Distribution: Central and southeastern China including Taiwan. Occurs chiefly in mountainous country near streams. Frequents woodland, scrub and semicultivated land.

Pope's Tree Viper, *Trimeresurus popeorum* Smith.

Identification: Separated from *T. stejnegeri* primarily by the structure of the male sexual organs; however, the following additional points of difference are noteworthy:

1. Iris yellow rather than reddish;
2. Size larger, reaching about 3 feet;
3. Lateral stripe indistinct in adult.

Distribution: Assam and Burma east to Cambodia and south through Malaysia and Indonesia. Inhabits hills between 3,000 and 5,000 feet for the most part. Common on tea plantations.

White-lipped Tree Viper, *Trimeresurus alboiabris* Gray.

Identification: First upper lip shield fused with nasal shield; white lateral line in males only; upper lip pale green, yellow or white; green of body generally somewhat paler than in *T. stejnegeri;* iris of eye yellow; end of tail dark red.

Average length 15 to 25 inches, maximum 36 inches; females considerably larger than males.

Distribution: Northeastern India to southeastern China, including Taiwan and Hainan, thence south through the Sunda Archipelago. Frequents lightly wooded or brushy areas; common on hillsides but rare above 1,500 feet; often found about human habitations including suburban gardens.

Indian Green Tree Viper, *Trimeresurus gramineus* (Shaw).

Identification: Differs from the other Asian green pit vipers in that most of the dorsal scales are smooth, keels being present only on a few posterior rows.

Green usually with darker flecking; light lateral line irregular; end of tail greenish; iris of eye yellow (see plate IV, fig. 3).

Average length 25 to 30 inches; maximum 40 inches.

Distribution: Peninsular India, chiefly in hilly country with dense undergrowth.

Remarks: The habits of these arboreal green vipers appear to be much the same. All are chiefly active at night remaining coiled in vegetation or hidden under bark or other cover during the day. They usually remain quiet when approached, but often strike if touched or otherwise threatened. They are reported to be a leading cause of snakebite accidents in Taiwan, Java, and Thailand. Persons picking tea, cutting bamboo, or clearing undergrowth are most often injured. Fatalities are unknown among adults, but have been reported in children.

An antivenin against *"Trimeresurus gramineus"* venom was produced by the Taiwan Serum Vaccine Laboratory, Taipei. Since true *T. gramineus* does not occur on Taiwan, the antivenin was probably for use against the venom of *T. stejnegeri*, the common green tree viper of the island.

So far as known, the tree vipers are live-bearing; there are 6 to 25 young in a litter.

Mangrove Viper, *Trimeresurus purpureomaculatus* (Gray).

Identification: General body build about the same as that of the green tree vipers; usually 25 or 27 scale rows at midbody vs. 19 or 21 in the green vipers.

Color variable—one common variety purplish brown with or without a whitish lateral line and with or without green spots. Another color phase is olive or gray irregularly spotted with brown. Tail uniformly brown or spotted gray and brown; belly white more or less clouded with brown.

Average length 30 to 35 inches, maximum about 40 inches.

Distribution: East Bengal, southern Burma, Malay Peninsula, Sumatra and Andaman islands. Largely restricted to the seacoast and to islands; particularly common in mangrove swamps.

Remarks: Usually found in low vegetation or among rocks. A fairly common cause of snakebite in coastal Malaya, but fatalities have not been recorded. There is no antivenin against the venom of this snake.

Sea snakes present in this region are discussed in Chapter VIII.

REFERENCES

BOURRET, Rene. 1936. Les Serpents de l'Indochine. H. Basuyau: Toulouse, 2 vols. vol. I, 141 pp., 14 figs.; vol. II, 505 pp., 189 figs.

DERANIYAGALA, P. E. P. 1960. The Taxonomy of the Cobras of Southeastern Asia. Spolia Zeylanica, vol. 29, pp. 41–63, fig. 1, pls. 1–4. 1961. The Taxonomy of the Cobras of Southeastern Asia, Part 2. *ibid.*, vol. 29, part 2, pp. 205–232, figs. 1–3, pls. 1–2.

HAAS, C. P. J. De. 1950. Checklist of the Snakes of the Indo-Australian Archipelago. Treubia, vol. 20, pp. 511–625.

HAILE, N. S. 1958. The Snakes of Borneo with a Key to the Species. Sarawak Mus. Jour., vol. 8 (12 n.s.), pp. 743–771, pls. 22–23, figs. a–h.

MINTON, Sherman A. Jr. 1966. A Contribution to the Herpetology of West Pakistan. Bull. Amer. Mus. Nat. Hist., vol. 134, art. 2, pp. 27–184, figs. 1–2, pls. 9–36.

ROMER, J. D. 1961. Annotated Checklist with Keys to the Snakes of Hong Kong. Mem. Hong Kong Nat. Hist. Soc. (5): pp. 1–14.

ROOIJ, Nelly De. 1917. The Reptiles of the Indo-Australian Archipelago, vol. II, Ophidia. E. J. Brill; Leiden. 334 pp., 117 figs.

SMITH, Malcolm A. 1943. The Fauna of British India including Ceylon and Burma. Reptilia and Amphibia, vol. 3, Serpentes. Taylor and Francis: London. 583 pp., 166 figs., map.

SWAN, Lawrence W. and Alan E. LEVITON. 1962. The Herpetology of Nepal: a History, Checklist and Zoogeographical Analysis of the Herpetofauna. Proc. California Acad. Sci., ser. 4, vol. 32, pp. 103–147, 4 figs.

TWEEDIE, M. W. F. 1954. The Snakes of Malaya. Government Printing Office: Singapore. 139 pp., 12 pls. 27 figs.

TAYLOR, Edward H. 1965. The Serpents of Thailand and Adjacent Waters. Univ. Kansas Sci. Bull., vol. 45, no. 9, pp. 609–1096, figs. 1–125, map.

WALL, Frank. 1928. The Poisonous Terrestrial Snakes of our British Indian Dominions (including Ceylon) and How to Recognize Them. Diocesan Press: Bombay. 173 pp.

Section 9

THE FAR EAST

Definition of the Region:
Includes the Philippines; Taiwan; the Ryukyu and Bonin Islands; Japan; Korea; Mongolia; Siberia; Russian Far East Asia and the Chinese provinces of Heilungkiang, Kirin, Inner Mongolia, Liaoning, Hopeh, Shantung, Shansi, Shensi, Ningsia Tui, Kansu, Hupeh, Anhwei, and Kiangsu.

TABLE OF CONTENTS

TABLE 13.—DISTRIBUTION OF POISONOUS SNAKES IN THE FAR EAST

	Philippines	Taiwan	Ryukyu Is.	Japan	E. China*	Korea	Mongolia	Siberia & Russian Far East Asia
ELAPIDAE								
Bungarus multicinctus		X						
Calliophis boettgeri			X					
C. calligaster	X							
C. iwasakii			X					
C. macclellandii		X						
C. sauteri		X						
Maticora intestinalis	X							
Naja naja	X	X			SE			
Ophiophagus hannah	X							
VIPERIDAE								
Vipera berus					N		X	X
V. russelii		X						
CROTALIDAE								
Agkistrodon acutus		X			X			
A. halys		?	?	X	X	X	X	X
Trimeresurus albolabris		?	S					
T. elegans			S					
T. flavomaculatus	X							
T. flavoviridis			X					
T. gracilis		X			X			
T. jerdonii					X			
T. monticola		X			S			
T. mucrosquamatus		X			S			
T. okinavensis			X					
T. stejnegeri		X			?			
T. wagleri	X							

*Hupeh, Anhwei, Kiangsu, Shensi, Shansi, Honan, Shantung, Hopeh, Liaoning, Kansu, Ningsia Hui, Kirin, Heilungkiang and Inner Mongolia.

Certain groups of adjoining nations or provinces are here treated as units. The symbol X indicates distribution of the species is widespread within the unit. Restriction of a species to part of a unit is indicated appropriately (SW = southwest, etc.). The symbol ? indicates suspected occurrence of a species within the area without valid literature.

INTRODUCTION

Zoogeographically, it is difficult to delimit or characterize the Far East. Insofar as the snake fauna goes, the southern part of this region closely resembles southeast Asia. There are archipelagoes (Philippines, Ryukyu) with more or less remote and diverse connections with the mainland. These have acted as secondary centers of evolution fostering development of distinctive island races of many snakes. Most of these races are sufficiently similar to mainland forms that they are not considered separately in this manual. The moist tropical climate that characterizes the southern part of the Far East excludes snakes requiring an arid or semiarid environment. Toward the north and inland, the snake fauna rapidly diminishes to a very few species because of the increasingly cold and dry climate.

Many areas in the Far East are densely populated and people live under conditions which expose them to snakebite. Many are engaged in farming and related occupations which may take

MAP. 10.—Section 9, the Far East.

them into the habitats of snakes. The incidence of snakebite is high in some localities, however the mortality is well below that reported in parts of India and Burma. The reasons for this are not altogether understood. The most important venomous snakes of the Far East are pit vipers, especially those of the genus *Trimeresurus*. Cobras are important toward the south. Sea snakes are numerous, but cases of serious sea snake bite are rare.

GENERIC AND SPECIES DESCRIPTIONS

ELAPIDAE: Genus *Bungarus* Daudin, 1803.

Kraits.

Twelve species are recognized; all inhabit the region of southeast Asia. Occasional individuals of *B. fasciatus* attain lengths of 7 feet. Most species are of moderate (4 to 5 feet) length, but all are considered extremely dangerous. A single species, *B. multicinctus*, is found in this region (see pp. 120–121).

Definition: Head small, flattened, slightly distinct from neck; no distinct canthus. Body moderately slender, cylindrical; tail short.

Eyes small; pupils round or vertically subelliptical.

Headscales: The usual 9 on the crown; frontal broad. Laterally, nasal in broad contact with single preocular.

Body scales: Dorsals smooth, vertebral row enlarged and hexagonal (strongly so except in *B. lividus*), in 13–17 oblique rows at midbody. Ventrals 193–237; anal plate entire; subcaudals single or paired (all paired only in some specimens of *B. bungaroides*), 23–56.

Maxillary teeth: Two large tubular fangs with external grooves followed, after an interspace, by 1–4 small, feebly-grooved teeth.

ELAPIDAE: Genus *Calliophis* Gray, 1834.

Oriental coral snakes.

Thirteen species are recognized; all inhabit the region of southeastern Asia. Most are small species but a few exceed 3 feet in length. At least the larger individuals are considered dangerous. Five species are known from this region (see pp. 121–122).

Definition: Head small, not distinct from body. Body cylindrical, slender and elongated; tail short.

Eyes small to moderate in size; pupils round.

Head scales: The normal 9 on the crown; rostral broad and rounded, no canthus. Laterally, nasal in contact with single preocular or separated from it by prefrontal; preocular absent in *C. bibroni*.

Body scales: Dorsals smooth, in 13–15 nonoblique rows throughout body. Ventrals 190–320; anal plate entire or divided.; subcaudals usually paired, occasionally single in *C. macclellandii* 12–44.

Maxillary teeth: Two large tubular fangs with ex-

KEY TO GENERA

1. A. Loreal pit present_____ 2
   B. Loreal pit absent_____ 3
2. A. No enlarged crown shields_____ *Trimeresurus*
   B. Enlarged crown shields present_____ _____ *Agkistrodon*
3. A. Enlarged crown shields absent or reduced in
      number_____ 4
   B. Eight or 9 enlarged crown shields_____ 5
4. A. Ventrals extending full width of belly_____ *Vipera*
   B. Ventrals not extending full width of belly or
      absent_____ NP*
5. A. Tail paddle-shaped_____ Sea snakes
      see Chapter VIII
   B. Tail not paddle-shaped_____ 6
6. A. Loreal scale present_____ NP
   B. Loreal scale absent_____ 7
7. A. Dorsal scales smooth_____ 8
   B. Dorsal scales keeled_____ NP
8. A. Vertebral scale row enlarged; subcaudals single_____ *Bungarus*
   B. Not as above_____ 9
9. A. Body scales in 17 or more rows on neck; hood
      seen in life_____ 10
   B. Body scales in 13 or 15 rows on neck; no hood_____ 11
10. A. Occipital shields present; anterior subcaudals
      single_____ *Ophiophagus*
    B. Occipital shields absent; subcaudals paired_____ *Naja*
11. A. Venom glands in normal position; anal shield
      usually divided_____ *Calliophis*
    B. Venom glands extended well back into body;
      anal entire_____ *Maticora*

---

* N.P.—Nonpoisonous

ternal grooves followed, after an interspace, by 0–3 small teeth.

ELAPIDAE: Genus *Maticora* Gray, 1834.

Long-glanded coral snakes.

Two species are found in the region of southeastern Asia: from Thailand and the Philippines to Sumatra, Java, Borneo, and Celebes. These snakes are relatively small and slender but individuals of one species, *M. bivirgata*, occasionally approach 5 feet in length; such individuals are believed to be capable of inflicting a dangerous bite (see p. 122, fig. 84). One species, *M. intestinalis*, inhabits the Philippines.

Definition: Head small and not distinct from body. Body cylindrical, slender and elongated; tail short.

Eyes small to moderate; pupils round.

Head scales: The usual 9 on the crown; no canthus; rostral broad and rounded. Laterally, nasal in broad contact with single preocular; eye in contact with supralabial row.

Body scales: Dorsals smooth, in 13 nonoblique rows

throughout body. Ventrals 197–293; anal plate entire; subcaudals paired, 15–50.

Maxillary teeth: Two large tubular fangs; no other teeth on the bone.

Remarks: The only consistent difference between these snakes and those of the genus *Calliophis* is that *Maticora* has elongated venom glands that extend posteriorly for about one-third of the body length. The heart has been pushed back to the middle third of the body, where it can be felt (in preserved specimens) as a hard object, thus identifying the genus.

ELAPIDAE: Genus *Naja* Laurenti, 1768.

Cobras.

Six species are recognized; all are African except the Asiatic cobra, *Naja naja*, and range throughout the African continent except for the drifting sand areas of the Sahara region. They are snakes of moderate (4 feet) to large (8 feet) size, with large fangs and toxic venom. The species, *N. nigricollis*, "spits" its venom at the eyes of an aggressor; it is found in the southern part of the region of north Africa. The Egyptian cobra (*Naja haje*) and the western subspecies of the Asiatic cobra (*Naja naja oxiana*) are found in the Near and Middle East region. *N. naja* is the only species in this region (see pp. 123–125).

Definition: Head rather broad, flattened, only slightly distinct from neck; snout rounded, a distinct canthus. Body moderately slender, slightly depressed, tapered; neck capable of expansion into hood; tail of moderate length.

Eyes moderate in size; pupils round.

Head scales: The usual 9 on the crown; frontal short; rostral rounded. Laterally, nasal in contact with the one or two preoculars.

Body scales: Dorsals smooth, in 17–25 oblique rows at midbody, usually more on the neck, fewer posteriorly. Ventrals 159–232; anal plate entire; subcaudals 42–88, mostly paired.

Maxillary teeth: Two rather large tubular fangs with external grooves followed, after an interspace, by 0–3 small teeth.

ELAPIDAE: Genus *Ophiophagus* Günther, 1864

King cobra.

A single species, *O. hannah*, is found in southeastern Asia and the Philippines. It attains a length of 16 to 18 feet, and is considered one of the world's most dangerous snakes (see pp. 125–126).

Definition: Head relatively short, flattened, moderately distinct from neck; snout broad, rounded, canthus indistinct. Body slender, tapering, neck region capable of expanding into small hood; tail long.

Eyes moderate in size; pupils round.

Head scales: The usual 9 on the crown, plus a pair of large occipitals in contact with one another behind the parietals. Laterally, nasal in narrow contact with elongate preocular.

Body scales: Dorsals smooth, in 15 oblique rows at mibody and posteriorly, more (17–19) on neck. Ventrals 240–254; subcaudals 84–104, the anterior ones single, the remainder paired.

Maxillary teeth: Two relatively short fangs (about ½ inch in a large specimen) with external grooves followed, after an interspace, by 3 small teeth.

VIPERIDAE: Genus *Vipera* Laurenti, 1768.

True adders.

Eleven species are recognized. This is an especially variable group, with some members that are small and relatively innocuous (e.g., *V. berus*, which is found in this region. See p. 74.) and others that are extremely dangerous (*V. lebetina*, *V. russelii*). They are found from northern Eurasia throughout that continent and into north Africa. One species ranges into the East Indies (*V. russelii*), as well as into the southern part of this region (see p. 127).

Definition: Head broad, distinct from narrow neck;

canthus distinct. Body cylindrical, varying from moderately slender to stout; tail short.

Eyes moderate in size to small; pupils vertically elliptical.

Head scales: Variable: one species (*V. ursinii*) has all 9 crown scutes, most species have at least the supraoculars, but even these are absent in one (*V. lebetina*): head otherwise covered with small scales. Laterally, nasal in contact with rostral or separated by a single enlarged scale (the nasorostral) eye separated from supralabials by 1–4 rows of small scales.

Body scales: Dorsals keeled, with apical pits, in 19–31 nonoblique rows at midbody. Ventrals rounded, 120–180; subcaudals paired, 20–64.

CROTALIDAE: Genus *Agkistrodon* Beauvois, 1799.

Moccasins and Asian pit vipers.

Twelve species are recognized. Three of these are in North and Central America; the others are in Asia, with one species, *A. halys*, ranging westward to southeastern Europe. The American copperhead (*A. contortrix*) and the Eurasian mamushi and its relatives (*A. halys*) seldom inflict a serious bite but *A. acutus* and *A. rhodostoma* of southeastern Asia, as well as the cottonmouth (*A. piscivorus*) of the southeastern United States, are dangerous species.

Definition: Head broad, flattened, very distinct from narrow neck; a sharply-distinguished canthus. Body cylindrical or depressed, tapered, moderately stout to stout; tail short to moderately long.

Eyes moderate in size; pupils vertically elliptical.

Head scales: The usual 9 on the crown in most species; internasals and prefrontals broken up into small scales in some Asian forms; a pointed nasal appendage in some. Laterally, loreal pit separated from labials or its anterior border formed by second supralabial. Loreal scale present or absent.

Body scales: Dorsals smooth (in *A. rhodostoma* only) or keeled, with apical pits, in 17–27 nonoblique

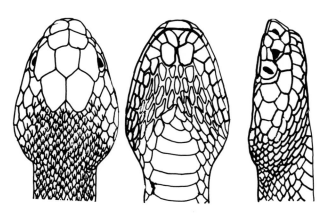

FIGURE 93.—Head Scales of *Agkistrodon halys*. Note presence of enlarged head scutes and loreal pit, characteristic of this genus. Redrawn from Maki, 1931.

rows. Ventrals 125–174; subcaudals single anteriorly or paired throughout, 21–68.

Sharp-nosed Pit Viper, *Agkistrodon acutus* (Günther).

Identification: A pit viper with the snout ending in an upturned pointed appendage and large shields on the crown.

Ground color gray or brown with dark brown crossbands narrow at the center of the back, wide on the sides, wide parts often tinged with dull orange; belly cream with large black spots that extend onto the sides; top of head dark brown, sides below eye yellow. The entire color scheme suggests that of the United States copperhead.

Average length 35 to 45 inches; maximum about 5 feet.

FIGURE 94.—Sharp-nosed Pit Viper, *Agkistrodon acutus*. The most dangerous pit viper of the Far East. Photo by New York Zoological Society. (See also plate V, figure 1.)

Distribution: Southern China, northern Viet Nam, Hainan, Taiwan. Found mostly in rocky, wooded, hilly country.

Remarks: A sedentary snake but alert and irritable; it strikes without hesitation when alarmed. Data from Taiwan indicate it is the most dangerous pit viper of the Far East. Antivenin is produced by the Taiwan Serum Vaccine Laboratory, Taipei.

Mamushi, *Agkistrodon halys* (Pallas).

Identification: Over much of its range the facial pit alone suffices to identify this snake. Presence of large crown shields distinguish it from other pit vipers within its range.

Yellowish or reddish brown with wide dark brown crossbands, irregular in outline and margined with black; side of head above eye dark brown or black, below eye pale buff to white; belly white or cream with black blotches. While the pattern and colors of the sharp-nosed pit viper suggest those of the copperhead to an American herpetologist, the mamushi suggests the

FIGURE 95.—Korean Mamushi, *Agkistrodon halys brevicaudus*. The Japanese Mamushi, *A. halys blomhoffii*, is similar but has fewer blotches. Photo by New York Zoological Society. (See also plate V, figure 5.)

cottonmouth moccasin. The resemblance is probably not coincidental; the American snakes very likely evolved from ancestors that migrated across a land bridge from Asia.

Average length of the mamushi is 20 to 26 inches; maximum about 35 inches.

The above account is confined to the subspecies *Agkistrodon halys blomohoffii* and *A. h. brevicaudus*.

Distribution: Japan and the Bonin and Pescadores Islands; Korea; and eastern and northern China. It evidently occurs in a wide variety of environments from low marshy river valleys to mountains at elevations up to 12,000 feet. It is occasionally found in the environs of Tokyo and other large cities.

Remarks: Generally an inoffensive diurnal snake that seeks to escape whenever possible. It flattens its body and vibrates its tail when angry. Despite its mild disposition, some 2,000 to 3,000 snakebites are reported annually in Japan. Fatalities are known but are most exceptional—about 1 per 1,000 bites. Woodcutters and farmers are most often bitten. Eight of 65 patients treated at a metropolitan hospital were bitten while preparing snakes for the table or for "Mamushi Whiskey," a concoction probably more deadly than the snake that goes into its manufacture.

Antivenin is produced by the Institute for Medical Science, Tokyo.

CROTALIDAE: Genus *Trimeresurus* Lacépède, 1804.

Asian lance-headed vipers.

About 30 species are currently recognized. All are found in southeast Asia and the adjacent island chains. The large species are dangerous; many of the smaller kinds can deliver a venomous bite which is very painful, but seldom if ever fatal.

There are 3 general groups of these snakes:

1. Large, long-bodied and long-tailed terrestrial snakes that are often brightly-colored with contrasting patterns;

2. Small short-bodied and short-tailed terrestrial

snakes, commonly with dull patterns of brown blotches;

3. Small, moderately long-bodied arboreal snakes with prehensile tail, body coloration tending toward unicolor greens, light browns, or light speckles.

Definition: Head broad, flattened, very distinct from narrow neck; canthus obtuse to sharp. Body cylindrical to moderately compressed, moderately slender to stout; tail short to moderately long.

Eyes small to moderate in size; pupils vertically elliptical.

Head scales: Supraoculars present, a pair of internasals often present; remainder of crown covered with small scales. Laterally, a nasal pore in prenasal, 2 enlarged preoculars, eye separated from supralabials by 1–4 rows of small scales.

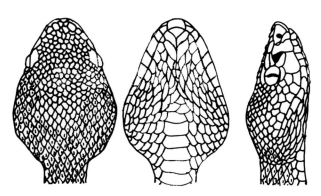

FIGURE 96.—Head Scales of the Okinawa Habu, *Trimeresurus flavoviridis*. Note absence of most crown scutes, characteristic of this genus. Redrawn from Maki, 1931.

Body scales: Dorsals feebly to strongly keeled, in 13–37 nonoblique rows. Ventrals 129–231; subcaudals paired, 21–92.

Okinawa Habu, *Trimeresurus flavoviridis* (Hallowell).

Identification: One of the Asian lance-headed pit vipers; head large, crown with small scales; body slender, gracefully proportioned, tail not prehensile. Scales around midbody 33–37; ventrals 222–231.

Ground color light olive or brown with elongated dark green or brownish blotches edged with yellow and sometimes enclosing yellow spots; the blotches often fuse to produce wavy stripes; underside whitish with dark mottling along the edges.

Average length 4 to 5 feet; maximum 7½ feet; it is the largest of the Asian lance-heads.

Distribution: Restricted to the Amami and Okinawa islands where it is common on the larger islands of volcanic origin, but is never seen on the smaller coral islands. It is most frequently found in the transition zone between cultivated fields and palm forest, living in rock walls, old tombs, and caves.

Remarks: An active, mostly nocturnal snake that

FIGURE 97.—Okinawa Habu, *Trimeresurus flavoviridis.* Photo by Robert E. Kuntz. (See also plate IV, figure 5.)

frequently enters dwellings and other man-made structures probably in search of rats and mice. It is a bold and irritable reptile striking with great rapidity and long reach. In the Amami islands the incidence of snakebite is very high—about 2 per 1,000 population. Fortunately, habu venom is of low toxicity and only about 3 percent of the bites are fatal; however, another 6 to 8 percent have permanent disability as a result of the bite.

Antivenin against *T. flavoviridis* venom is produced by the Institute for Medical Science, Tokyo, and the Laboratory for Chemotherapy and Serum Therapy, Kumamoto, Japan.

The habu is one of the comparatively few pit vipers that lays eggs.

The himehabu (*Trimeresurus okinavensis*) has much the same distribution as the habu but is a smaller, heavier snake. It is sluggish and rarely causes snakebite. The Sakishima habu (*T. elegans*) (plate IV, fig. 6) is a smaller version of the Okinawa habu and occurs in the southern Ryukyus. It has 182–191 ventrals and 66–77 subcaudals.

FIGURE 98.—Himehabu, *Trimeresurus okinavensis.* Photo by Robert E. Kuntz. (See also plate IV, figure 6.)

Chinese Habu, *Trimeresurus mucrosquamatus* (Cantor).

Identification: Very similar to the habus of the Ryukyus; has 198–219 ventrals and 76–96 subcaudals; 23–27 scale rows at midbody.

Grayish brown to buff or olive; three rows of darker gray or brown spots with narrow yellow edges; those of middle row largest, occasionally fused to form a broken wavy stripe; belly whitish suffused with brown (see plate V, fig. 4).

Average length 30 to 40 inches; maximum about 4 feet.

Distribution: Taiwan and southern China west through northern Viet Nam and Laos to eastern Burma. Commonest in hilly areas with grass or sparse forest but at no great altitude. Often found in suburban and agricultural districts.

Remarks: Most of what has been said of the Okinawa habu appears to be true of this snake also. It is often implicated in snakebite accidents. Antivenin is produced by the Taiwan Serum Vaccine Laboratory, Taipei.

Chinese Mountain Viper, *Trimeresurus monticola* Günther.

Identification: A pit viper of decidedly stockier build than the Chinese habu and lacking the green color and prehensile tail of the tree vipers. Usually 23–27 scale rows at midbody; fewer than 200 ventrals; 65 or fewer subcaudals.

Gray or olive speckled with black; series of large squarish brown or reddish blotches; top of head dark brown or black sometimes with light Y-shaped mark; belly white mottled with dark brown.

Average length 3 to 3½ feet, maximum about 4 feet.

FIGURE 99.—Chinese Mountain Viper, *Trimeresurus monticola.* Photo by New York Zoological Society. (See also plate V, figure 3.)

Distribution: Nepal eastward across mainland China and south through the Malay Peninsula. Usually found in wooded mountainous country to elevations of about 8,000 feet.

Remarks: Lays eggs that are guarded by the mother. These nesting females are said to be somewhat sullen and irritable; otherwise it is a placid sluggish snake. No antivenin is available.

Wagler's Pit Viper, *Trimeresurus wagleri* (Boie).

Identification: Stout with unusually wide head and prehensile tail; scales between eyes and on chin and throat strongly keeled.

Adults green with black-edged scales or black with scattered green spots; broad crossbands, green above shading to yellow on sides; head black above, sides yellow or greenish; belly greenish mottled with yellow; tail black. Young, green with a regular row of spots, each one half white and half red; tail reddish. There is a good deal of color variation, especially in the Philip-

FIGURE 100.—Wagler's Pit Viper, *Trimeresurus wagleri.* Photo by New York Zoological Society. (See also plate IV, figure 4.)

pines. Some populations are almost uniform green, others tend to retain the juvenile pattern. The keeled throat scales are diagnostic throughout the range.

Average length 30 to 35 inches; maximum about 40 inches.

Distribution: Thailand, Malaysia, Indonesia, Borneo and the Philippines. A common snake of lowland jungle and plantations.

Remarks: A tree snake of remarkably sluggish and gentle disposition at least during the day. It is sometimes kept unconfined in temples or tolerated about dwellings as an omen of good luck. The venom is fairly toxic for animals and present in good quantity, so the snake is capable of inflicting a dangerous bite. No specific antivenin is available.

The sea snakes are discussed in Chapter VIII.

REFERENCES

KUNTZ, Robert E. 1963. Snakes of Taiwan. U.S. Naval Medical Research Unit No. 2: Taipei, Taiwan, pp. 1–79, color pls., text figs. (not numbered).

LEVITON, Alan E. 1961. Keys to the Dangerously Venomous Terrestrial Snakes of the Philippine Islands. Silliman Jour., vol. 8, pp. 98–106, figs. 1–2.

POPE, Clifford. 1935. The Reptiles of China. Nat. Hist. Cent. Asia, vol. 10 New York, pp. i–iii + 1–604, 27 pls. figs., map.

STEJNEGER, L. 1907. Herpetology of Japan and Adjacent Territory. Bull. U.S. Nat. Mus., vol. 58, pp. i–xx, 1–577, 35 pls.

WERLER, John E. and Hugh L. KEEGAN. 1963. Venomous Snakes of the Pacific Area. *In* Venomous and Poisonous Animals and Noxious Plants of the Pacific Region. (H. L. Keegan and W. V. MacFarlane, eds.) Pergamon Press: Oxford, pp. 219–325, figs. 1–78.

Section 10

AUSTRALIA AND PACIFIC ISLANDS

Definition of the Region:

Includes the continent of Australia and the islands of Oceania east of Japan, the Ryukyus, and the Philippines and east of a line drawn between Timor (including the offshore island of Mòa) and the Tanimbar Islands, and between Celebes and the islands of Buru and Halmahera.

TABLE OF CONTENTS

TABLE 14.—DISTRIBUTION OF POISONOUS SNAKES OF AUSTRALIA & THE PACIFIC ISLANDS

	Obi	Ceram	Tanimber (Timor Laut)	Kai	New Guinea	Schouten	Aru	Fredrik-Hendrik	Melville	Australia	Bass Straits Islands	Tasmania	Bismarck Is.	Solomon Is.	Fiji Islands
ELAPIDAE															
Acanthophis antarcticus	X	X	X	X	X	X	X	?	X	X			W		
A. pyrrhus										C					
Apistocalamus grandis					X										
A. lamingtoni					X										
A. loennbergii					X										
A. loriae					X										
A. pratti					E										
Aspidomorphus christieanus															
A. diadema										N					
A. harriettae										S					
A. krefftii										NE					
A. minutus										E					
A. mnellerii					X					SE					
A. schlegelii		X			X	X							X		
A. squamulosus															
Brachyaspis curta										E					
Brachyurophis australis										SW					
B. campbelli										E					
B. fasciolatus										NE					
B. roperi										SW					
B. semifasciatus										N					
B. warro										SW					
B. woodjonesii										NE					
Demansia acutirostris										NE					
D. guttata										S					
D. modesta										NE					
D. olivacea					SE				X	W					
D. psammophis					SE					N					
D. textilis										X					

(Continued on page 142)

INTRODUCTION

Most of the islands of the Pacific Ocean have no poisonous land snakes, although those in equatorial waters are likely to have poisonous sea snakes just offshore (See Chapter VIII). In addition, some of the poisonous snakes that do occur on islands are so small that they cannot be considered a hazard to man.

Australia and New Guinea have large numbers of dangerously poisonous snakes but of the islands east of New Guinea only the Solomons have poisonous snakes which can even remotely be considered dangerous. The Fiji Islands, for example, have a poisonous snake (*Ogmodon vitianus*) but it is so small (15 inches) that its killing power is limited to the small animals on which it feeds. Further, it is so rare that most island residents have never seen it!

Australia, on the other hand, is the only continent which has more kinds of poisonous than nonpoisonous snakes. More than 60 percent of Australian snakes are poisonous and some are highly dangerous. Yet of the 60 species of poisonous land snakes that inhabit Australia, only about 16 are considered to be dangerous to an adult man. Several of these have rather restricted ranges and are not found in areas of high human population. For a country with such a high number of poisonous snakes, Australia has amazingly few deaths from snakebite—the annual rate being estimated at 1 in 2,000,000.

In New Guinea, just to the north, fewer than 25 percent of its snakes are poisonous. Of the 16 species of poisonous snakes, only 6 are considered highly dangerous and 4 of these are restricted to the southeastern coast adjacent to Australia. Thus, only 2 species, the death adder and the ikaheka snake, are of concern elsewhere on the island. The remainder of the poisonous species outside eastern New Guinea is made up of small burrowing snakes or species resembling whip snakes whose bites are of minor consequence.

Aside from sea snakes which are found offshore and in some of the rivers and lakes (see Chapter VIII), the poisonous snake fauna of this region is made up entirely of members of the cobra family (Elapidae). Although they are all elapids, none is a *true* cobra; none has a cobra-type hood (though several flatten the neck—or

even the whole body) and none stands up straight in cobra fashion as a threat. Many of the dangerous snakes of this region resemble North American whip snakes and since they lack any special physical characteristic (such as the rattle of the rattlesnake or the facial pit of the pit viper) or any unusual behavioral features, they

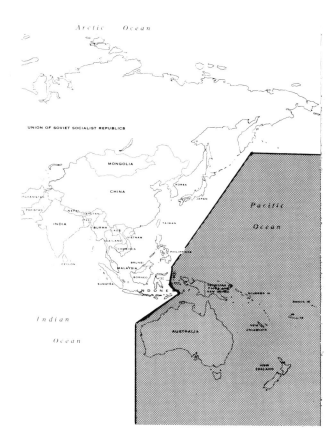

MAP 11.—Section 10, Australia and Pacific Islands. (Includes the Pacific Ocean eastward to the coasts of the Americas.)

are particularly difficult to distinguish from nonpoisonous species.

True, the dangerous death adder has the appearance of a viper (which it isn't), but other dangerous species look like harmless racers, rat snakes, or king snakes. About the only way to identify a poisonous snake from this region is to kill it and look for fangs. (Chapter III, fig. 5). Even this is not a foolproof method because some

TABLE 14.—DISTRIBUTION OF POISONOUS SNAKES OF AUSTRALIA & THE PACIFIC ISLANDS (continued)

	Obi	Ceram	Tanimber (Timor Laut)	Kai	New Guinea	Schouten	Aru	Fredrik-Hendrik	Melville	Australia	Bass Straits Islands	Tasmania	Bismarck Is.	Solomon Is.	Fiji Islands
ELAPIDAE (continued)															
Denisonia boschmai					S										
D. brunnea										S					
D. carpentariae										NE					
D. coronata										SW					
D. coronoides										S		X			
D. damelii										E					
D. devisi										E					
D. dwyeri										E					
D. fasciata										SW					
D. flagellum										SE					
D. gouldii										S					
D. maculata										NE					
D. nigrostriata										NE					
D. pallidiceps										NE					
D. par														X	
D. punctata										N					
D. ramsayi [1]										SE					
D. signata										E					
D. superba										SE		X			
D. suta										X					
Elapognathus minor										SW					
Glyphodon barnardi										NE					
G. dunmalli										NE					
G. tristis					SE		?			NE					
Hoplocephalus bitorquatus										NE					
H. bungaroides										SE					
H. stephensii										SE					
Micropechis elapoides							X							X	
M. ikaheka					X										

ELAPIDAE (continued)

Species						
Notechis scutatus		X	S			X
Ogmodon vitianus					X	X
Oxyuranus scutellatus	SE		NE	X		
Parademansia microlepidota [2]			SE			
Parapistocalamus hedigeri						X
Pseudapistocalamus nymanni [3]	X					
Pseudechis australis	SE		N	X		
P. colletti	SE		NE			
P. papuanus		X	SE			
P. porphyriacus			E, S			
Rhinoplocephalus bicolor			W			
Rhynchoelaps approximans			NW			
R. bertholdi			SW			
Toxicocalamus longissimus	E					
T. stanleyanus	X					
Tropidechis carinatus			E			
T. dunensis [4]			NE			
Ultrocalamus preussi	X					
Vermicella annulata			X			
V. bimaculata			SW			
V. calonota			SW			
V. minima			NW			
V. multifasciata [5]			N			

The symbol X indicates widespread distribution of the species within the area. The symbol ? indicates suspected presence of a species within the area without valid literature.

[1] Based on juvenile D. superba according to Worrell.
[2] Recently shown to be based on young individuals of Oxyuranus.
[3] Appears to be synonym of Apistocalamus pratti.
[4] Shown to be based on misidentified specimen of Dasypeltis.
[5] A synonym of V. annulata according to Worrell.

143

species have very short fangs that may be difficult to distinguish from the other teeth unless a microscope is at hand.

Elapid snakes typically lack a scale on the side of the face (loreal) which most colubrid snakes have. This means that only 2 scales lie between the nostril and the eye instead of the 3 that are seen in most harmless snakes. (A few kinds, e.g., *Glyphodon*, on rare occasions have a vertical suture through the preocular, forming a "loreal.") Any snake that lacks this scale should be viewed with suspicion. Fortunately, too, a rather large proportion of the harmless snakes of this region are pythons, boas, blind snakes, or highly specialized water snakes. These are much easier to distinguish from elapids than are the colubrids, which are the most common snakes elsewhere (See Chapter VI).

It is worth remembering, too, that almost all of the dangerously poisonous kinds of snakes in this region live on the ground. Only the Australian broad-headed snakes (*Hoplocephalus*), among poisonous species, are adapted for life in the trees and they are not considered highly dangerous. Otherwise, only the tiger snake of southern Australia has been reported to climb into low bushes. There are no highly specialized dangerous tree snakes such as the mambas of Africa or the tree vipers of southeast Asia and tropical America. There are many species of burrowing elapids in Australia, but none appears to be highly dangerous.

KEY TO GENERA

1. A. Tail paddle-shaped_____ 2
   B. Tail not paddle-shaped_____ 3
2. A. No enlarged crown shields_____ NP*
   B. At least some enlarged crown shields_____ Sea snakes
   (*see* Chapter VIII)
3. A. Dorsal scales smooth; no trace of a keel_____ 6
   B. Dorsal scales with a distinct keel_____ 4
4. A. Eye separated from upper labials by a row of
       small subocular scales; pupil elliptical_____ *Acanthophis*
   B. Eye touching upper labials; pupil round_____ 5
5. A. Dorsal scales roughly keeled; fewer than 220
       ventrals_____ *Tropidechis*
   B. Dorsal scales with a low keel; more than 220
       ventrals_____ *Oxyuranus*
6. A. Ventrals extend full width of belly_____ 7
   B. Ventrals extend half the width of the belly or less_____ NP
7. A. Ventrals with a lateral keel and notch_____ *Hoplocephalus*
   B. Ventrals rounded; no keel or notch_____ 8
8. A. Eye with a round pupil_____ 12
   B. Eye with an elliptical pupil_____ 9
9. A. Only six lower labials_____ *Apistocalamus*
   B. Seven or more lower labials_____ 10
10. A. Anal plate divided; subcaudals paired throughout____ *Aspidomorphus*
    B. Anal plate entire; at least some subcaudals single_____ 11
11. A. More than 18 rows of dorsals, slightly oblique at
        midbody_____ *Brachyaspis*
    B. Fewer than 18 rows of dorsals, not oblique at
        midbody_____ *Denisonia*

---

*NP—Nonpoisonous

KEY TO GENERA (continued)

12. A. Eye small; its length considerably less than its distance from lip_____ 18
    B. Eye moderate to large, its length about equal to or more than distance to lip_____ 13
13. A. First row of dorsals conspicuously broader than adjacent row; dorsal count 17 posteriorly, 17 or 19 at midbody_____ *Pseudechis*
    B. First row not conspicuously broader; if dorsal count 17 posteriorly, more than 19 at midbody_____ 14
14. A. At least some of subcaudals paired_____ 17
    B. All of subcaudals single_____ 15
15. A. Body very short (fewer than 150 ventrals) and rather stout_____ *Elapognathus*
    B. Body moderately long (more than 150 ventrals) and rather slender_____ 16
16. A. Frontal long, 1.5 to 2 times longer than broad; dorsals not oblique_____ *Denisonia*
    B. Frontal short, almost as wide as long; dorsals distinctly oblique_____ *Notechis*
17. A. More than 227 ventrals_____ *Oxyuranus*
    B. Fewer than 228 ventrals_____ *Demansia*
18. A. Body moderate to slender (fewer than 227 ventrals)_____ 21
    B. Body exceedingly long and slender (more than 226 ventrals)_____ 19
19. A. A preocular present; 2 to 3 small teeth following fangs after an interspace (Australia)_____ *Vermicella*
    B. No preocular; 4 to 5 teeth of decreasing size following fang without an interspace (New Guinea)_____ 20
20. A. A long terminal spine which is keeled above; internasals distinct_____ *Toxicocalamus*
    B. Terminal spine obtuse, not keeled; internasals fused with prefrontals_____ *Ultrocalamus*
21. A. Nasal in contact with preocular_____ 24
    B. Nasal separated from preocular (if present) by prefrontal_____ 22
22. A. Fewer than 156 ventrals (Fiji)_____ *Ogmodon*
    B. More than 155 ventrals (Australia, Solomons)_____ 23
23. A. Nasal barely touches 2nd labial_____ *Glyphodon*
    B. Nasal extends well over 2nd labial_____ *Parapistocalamus*
24. A. Rostral broad and free at sides; subcaudals single__ *Rhinoplocephalus*
    B. Rostral not free at sides; subcaudals paired_____ 25
25. A. Tail very short, fewer than 31 subcaudals (Australia)_____ 26
    B. Tail longer, more than 30 subcaudals (New Guinea and the Solomon Islands)_____ *Micropechis*
26. A. Rostral large and shovel-shaped, with a sharp edge____ *Brachyurophis*
    B. Rostral without a sharp edge_____ 27
27. A. Nasal in contact with first three upper labials_____ *Rhynchoelaps*
    B. Nasal in contact with first two labials only_____ *Vermicella*

GENERIC AND SPECIES DESCRIPTIONS

ELAPIDAE: Genus *Acanthophis* Daudin, 1803.

Death adders.

Two species are currently recognized. One of these, *A. antarcticus*, ranges widely over the region; the other, *A. pyrrhus*, is little known and is restricted to the desert areas of central and western Australia. *A. antarcticus* is one of the most deadly as well as one of the most widespread snakes of the region.

Definition: Head broad, flattened, and distinct from neck; a distinct canthus rostralis. Body thick and depressed; tail short with a long terminal spine.

Eyes moderate in size; pupils vertically elliptical.

Head scales: The usual 9 on crown, somewhat roughened with raised edges; supraoculars broad, overhanging eye. Eye separated from supralabials by a row of small suboculars.

Body scales: Dorsals distinctly keeled and pointed, in 21–23 rows at midbody. Ventrals 113–135; anal plate entire; subcaudals mostly single, some terminal ones paired, 40–52; a terminal spine made up of several scales.

Maxillary teeth: Two long recurved fangs followed, after an interspace, by 2–3 small teeth.

Death Adder, *Acanthophis antarcticus* (Shaw).

Identification: Extremely viperlike in appearance. Average length 18 to 24 inches; record is 36 inches.

Body color gray, brown, reddish, or yellowish with a more or less distinct pattern of irregular narrow dark crossbands. A pair of diverging dark markings on top of head. The long spine at the end of the tail is light yellowish or flesh-colored.

Distribution: Found throughout Australia except for the central desert regions, on Melville Island and New Guinea, and on the nearby islands of Aru, Ceram, Haruku, Kei, Obi, the Southern Islands, and Tanimbar. In Australia it usually inhabits dry scrub areas but has been found also in rain forest regions in Ceram and New Guinea.

Remarks: The death adder is active mainly at night and tends to be sluggish during the day. It often conceals itself in sand or dust and generally defends itself rather than retreat from such concealment. When disturbed it flattens the entire body and strikes out with viperlike speed. Although its fangs are short as compared with those of a viper, they are quite long for an elapid. It is an extremely dangerous snake and without treatment with specific antivenin the mortality rate has averaged about 50 percent.

Antivenin ("Death Adder") is produced only by the Commonwealth Serum Laboratories of Australia.

ELAPIDAE: Genus *Apistocalamus* Boulenger, 1898.

Five species are recognized, all in New Guinea. They are small burrowing snakes with poorly-defined fangs. Only one (*A. grandis*) attains a length of over 2 feet; it is known to grow to 37 inches. None is believed to be highly dangerous, though any elapid more than 2 feet long should be treated with respect.

Definition: Head small, somewhat flattened, and not distinct from body; body slender; tail short with distinct terminal spine which has a dorsal keel.

Eyes very small; pupils vertically elliptical.

Head scales: The usual 9 on crown, supraoculars short, parietals long. Single preocular in contact with nasal or narrowly separated from it by second supralabial.

Body scales: Dorsals smooth in 15 rows throughout body. Ventrals 173–226; anal divided (entire in one species, *A. lamingtoni*); subcaudals usually paired (a few or all occasionally single), 22–59.

Maxillary teeth: Two small fangs followed, without an interspace, by 3–4 teeth that gradually decrease in length.

ELAPIDAE: Genus *Aspidomorphus* Fitzinger, 1843.

Crowned snakes.

Eight species are known. Two of these inhabit New Guinea and neighboring islands, the others are restricted to Australia. All are small; the largest attains a length of about 30 inches. None is considered dangerous to man.

Definition: Head flattened and distinct from neck; body moderately slender to stout; tail relatively short, without an elongated spine.

Eyes small; pupils vertically elliptical in most; round in *A. muellerii* (Schlegel).

FIGURE 101.—Death Adder, *Acanthophis antarcticus*. The most viperlike of all Australian elapid snakes. Photo by W. A. Pluemer, National Audubon Society.

Head scales: The usual 9 on crown, supraoculars long; preocular generally in contact with nasal.

Body scales: Dorsals smooth, in 15 rows throughout body or in 17 rows which may be reduced to 15 posteriorly. Ventrals 139–203; anal plate divided; subcaudals paired throughout, 25–62.

Maxillary teeth: Two large fangs followed, after a wide interspace, by 7–10 small teeth.

ELAPIDAE: Genus *Brachyaspis* Boulenger, 1896.

Bardick.

The single, little-known species (*B. curta*) is found in southwestern Australia. It is small, attaining a length of about 20 inches, and is capable of delivering a very painful, though not a lethal, bite.

Definition: Head large and distinct from the neck; an obtuse canthus rostralis. Body short and relatively stout; tail short.

Eyes small; pupils vertically elliptical.

Head scales: The usual 9 on the crown; frontal long and broader than supraoculars. Nasal usually in contact with preocular, but may be narrowly separated from it by prefrontal.

Body scales: Dorsals smooth in 19 slightly oblique rows at midbody reduced to 15 or 13 posteriorly. Ventrals, 128–138; anal plate entire; subcaudals single, 30–35.

Maxillary teeth: Two large fangs followed, after an interspace, by 2–5 small teeth.

ELAPIDAE: Genus *Brachyurophis* Günther, 1863.

Girdled snakes.

Seven species are currently recognized. They inhabit most of Australia except for the humid southeastern coastal regions. All are small sand-dwelling, burrowing species and are not believed to be dangerous.

Definition: Head short and not distinct from neck; snout distinctly pointed; no canthus rostralis. Body moderately slender with little taper; tail short.

Eyes small; pupils round.

Head scales: The usual 9 on the crown; rostral shovel-like with sharp anterior edge and with an angulate rear edge that partly divides internasals. Laterally, nasal in contact with preocular.

Body scales: Dorsals smooth, in 15–17 nonoblique rows at midbody. Ventrals 133–170; anal plate divided; subcaudals paired, 17–27.

Maxillary teeth: Two moderately large fangs with external groove followed, after an interspace, by a single small tooth.

ELAPIDAE: Genus *Demansia* Gray, 1842.

Brown snakes and whip snakes.

Six species are currently recognized, two of which are highly dangerous to man. Both *D. textilis* and *D.*

olivacea are found in southeastern New Guinea as well as on mainland Australia; the latter occurs also on Melville Island. The other species are restricted to mainland Australia.

Definition: Head elongate with a distinct canthus rostralis, only slightly distinct from neck. Body slender and racerlike; tail long and tapering.

Eyes large; pupils round.

Head scales: The usual 9 on the crown; frontal long and narrow. Laterally, nasal in contact with single preocular.

Body scales: Dorsals smooth, in 15–21 rows at midbody, more anteriorly and fewer posteriorly. Ventrals 167–225; anal plate divided; subcaudals paired throughout, 44–92.

Maxillary teeth: Two relatively short fangs followed, after an interspace, by 8–13 small teeth.

Black Whip Snake, *Demansia olivacea* (Gray).

Identification: This snake superficially resembles the harmless racers and whip snakes of North America and Eurasia. However, the short snout, with only two scales between nostril and eye, warns of its elapid relationship. Adults average 4 to 5 feet; occasional individuals exceed 6 feet.

Rich brown above, fading to a greenish-blue underneath. Each body scale edged with black; skin between scales with many irregular light stipple marks. A dark collar sometimes present; the entire coloration becomes darker toward the tail. Head sometimes spotted, with or without light markings on sides.

Dorsals in 15 rows at midbody; ventrals 180–200; subcaudals 69–105.

Distribution: Found in open sandy areas of northern Australia, southeastern New Guinea, and on Melville Island.

Remarks: The black whip snake is active during the day. It is fast-moving and normally inoffensive. Ordinarily it will flee if able. However, if injured or cornered it will defend itself fiercely and may inflict several bites in rapid succession. The bite of a large individual is presumed to be dangerous.

A polyvalent antivenin ("Brown Snake") is made for this group of snakes by the Commonwealth Serum Laboratories of Australia.

Australian Brown Snake, *Demansia textilis* (Duméril, Bibron and Duméril).

Identification: Head narrow and deep, slightly distinct from neck. Adult snakes average 5 to 6 feet; record length about 7 feet.

Body color almost any shade of brown, ranging from light grayish tan, through reddish brown, to dark brown. Juveniles may have a series of distinct narrow crossbands (about 35 on body, 15 on tail) plus a dark collar. Most adults almost unicolor above. Many have conspicuous dark spots or blotches on the cream, gray, or yellowish belly.

Dorsals in 17–19 rows at midbody; ventrals 184–225; subcaudals 45–75.

Distribution: Widely distributed through the drier areas of Australia, and in eastern New Guinea. Found in wheat fields and rice fields, and in some of the irrigated lands.

FIGURE 102.—Australian Brown Snake, *Demansia textilis.* This fast-moving snake is probably responsible for more deaths in Australia than any other snake. Photo by Eric Worrell.

Remarks: This is a fast-moving and agile snake that becomes aggressive if disturbed. When aroused it flattens its neck and raises it from the ground in an S-shaped loop. Large individuals should be treated with respect. Due to its common occurrence and toxic venom, it may be responsible for more deaths than any other Australian snake. It will strike repeatedly if antagonized.

An antivenin for this group of snakes ("Brown Snake") is produced by the Commonwealth Serum Laboratories of Australia. "Taipan" antivenin, also produced by Commonwealth, may be used also.

ELAPIDAE: Genus *Denisonia* Krefft, 1869.

Australian copperheads and ornamental snakes.

Nineteen species are recognized by Klemmer (1963: 290–294); except for a single species, they are all Australian. The interrelations of the snakes of this genus are not clear and Worrell (1963: 190, 192) does not believe that all belong to the same genus. Only 2 of the species appear to be highly dangerous. One of these is found in southeastern Australia and Tasmania (*D. superba*), the other (*D. par*) in the Solomon Islands.

Definition: Head small to moderate in size and not distinct or only slightly distinct from the neck. A fairly distinct canthus in some species, relatively indistinct in others. Body moderately stout to relatively slender; tail short.

Eyes moderate in size; pupils vertically elliptical in some, round in others.

Head scales: The typical 9 on the crown, frontal distinctly longer (1.5 to 2 times) than broad. Laterally, nasal in contact with preocular.

Body scales: Dorsals smooth, in 15–17 rows at midbody, fewer posteriorly. Ventrals 129–194; anal plate entire; subcaudals single in most species, paired in *D. par woodfordii* of the Solomons, 25–56.

Maxillary teeth: Two short fangs followed, after an interspace, by 3–10 small teeth.

Solomons Copperhead, *Denisonia par* (Boulenger).

Identification: Body moderately slender, slightly compressed; head somewhat flattened, distinct from neck. Adults average approximately 30 inches in length; record length a little more than 36 inches.

Body with a lustrous sheen. The color varies from sandy-brown through pink, reddish, and gray, to almost unicolor black. Faint, irregular crossbars may be visible but usually coloration is uniform with scale edges darker than centers.

Eyes moderate in size; pupils round or subelliptic.

Dorsals in 17 rows at midbody, reduced to 15 posteriorly. Ventrals 164–181; anal plate divided or entire; subcaudals 38–53, single (*D. p. par*) or paired (*D. par woodfordii*).

Distribution: Widespread in the Solomon Islands; not yet reported from Bougainville, Choiseul, or the islands south of Malaita and Guadalcanal. Found in rain forest, grasslands, and cocoanut plantations.

Remarks: Two other elapid snakes occur in the Solomons. *D. par* differs from *Parapistocalamus hedigeri* (so far known only from Bougainville) in having a longer tail (more than 37 subcaudals) and 2 postoculars; from *Micropechis elapoides* in having fewer ventrals (less than 185). See Williams and Parker (1964) for additional features.

This snake is considered potentially dangerous but nothing has been recorded on the effects of its bite.

Australian Copperhead, *Denisonia superba* (Günther).

Identification: Body moderately stout with short tail. Head flat and fairly broad, only slightly distinct from neck. Adult length averages 4 to 5 feet; record length (a Tasmanian specimen) about 6 feet.

Body color extremely variable. Coppery or reddish brown over much of its range; blackish or reddish with an obscure dark stripe down back in Blue Mountain region; a black back with yellowish or whitish sides in Bowral region. Sometimes (in Queensland) entirely black. Coloration of labial scales distinctive in Alpine specimens: each scale bicolored, upper and rear parts dark, lower and front parts edged with oblique dash of cream color.

Eyes with round pupil.

Dorsals in 15 rows at midbody; ventrals 145–160; 40–50 single subcaudals.

Distribution: Tasmania and the southern coastal region of Australia. Mainly an inhabitant of coastal mountain swamps; found in area from Victoria to New England ranges and at Mount Gambier and Kangaroo Island, South Australia.

Remarks: This is one of the best known of the venomous snakes of southern Australia. It is a dangerous but rather sluggish and inoffensive snake. It is unlikely to bite unless stepped on or picked up. Few bites are reported, but they have been serious.

"Tiger Snake" antivenin is commonly used for treatment of its bite. This is manufactured by the Commonwealth Serum Laboratories of Australia.

ELAPIDAE: Genus *Elapognathus* Boulenger, 1896.

Little brown snake.

The genus contains a single, little-known species (*E. minor*) that grows to a length of about 18 inches. It is found only in the southwestern section of Western Australia and is not considered dangerous.

Definition: Head small and only slightly distinct from neck. Body cylindrical and moderately stout; tail moderate in length.

Eyes rather large; pupils round.

Head scales: The usual 9 on the crown. Laterally, the long nasal is in contact with the single preocular.

Body scales: Dorsals smooth but finely striated, in 15 rows at midbody; fewer (13) posteriorly. Ventrals 120–130; anal plate entire; subcaudals single, 52–68.

Maxillary teeth: Two moderately large fangs; no other teeth on the bone.

ELAPIDAE: Genus *Glyphodon* Günther, 1858.

Australian collared snakes.

Three species are recognized. One (*G. tristis*) is found in southeastern New Guinea and some of the nearby islands in addition to the mainland of Australia; the others are restricted to Australia. One species grows to a length of 3 feet, but all reportedly refuse to bite even when teased, and are not considered dangerous.

Definition: Head small and slightly distinct from neck; no canthus rostralis. Body cylindrical and moderately slender; tail rather short.

Eyes small; pupils round.

Head scales: The usual 9 on the crown. Laterally, the prefrontal extends down to separate the nasal from the preocular.

Body scales: Dorsals smooth in 15–21 rows at midbody; species with 15 or 17 rows at midbody show no reduction posteriorly. Ventrals 163–190; anal plate divided; subcaudals paired (a few anterior ones single in a single specimen), 28–52.

Maxillary teeth: Two large fangs followed, after an interspace, by 6–10 small teeth.

ELAPIDAE: Genus *Hoplocephalus* Wagler, 1830.

Australian broad-headed snakes.

Three species are currently recognized; all are Australian. They appear to be the only Australian elapid snakes that are specialized for arboreal life. Only one of the species, *H. bungaroides*, attains a large enough size to be a danger, though probably the others also can deliver a painful bite.

Definition: Head broad and distinct from the slender neck; no canthus rostralis. Body relatively slender; tail moderately long.

Eyes moderate in size; pupils round.

Head scales: The normal 9 on the crown; frontal rather long. Laterally, nasal in contact with preocular.

Body scales: Dorsals smooth, in 21 rows at midbody, fewer posteriorly. Ventrals laterally angulate and notched (a typical indication of a treesnake), 210–227; anal plate entire; subcaudals single, 40–60.

Maxillary teeth: Two short fangs followed, after an interspace, by 4 small teeth.

Remarks: Three genera of harmless colubrid tree snakes also occur in Australia. All may be distinguished from *Hoplocephalus* by a loreal scale, (giving 3 scales between nostril and eye) and a longer tail (more than 80 subcaudals, all *paired*).

Australian Yellow-spotted Snake. *Hoplocephalus bungaroides* (Boie).

Identification: The broad head and eyes with round-pupils; angulated, keeled, and notched ventral scutes, and moderately long tail distinguish this snake. Adults average 3 to 4 feet in length; some individuals attain a length of 5 feet.

Ground color black or dark brown. Numerous conspicuous yellow spots form irregular crossbands or a broken network over the body. Tail solid black or almost so.

Dorsal scales in 21 rows at midbody; 214–221 ventrals; 40–60 subcaudals.

Distribution: Australia: the mountains and coastal regions of southern Queensland and New South Wales.

Remarks: This snake is active mainly at night. It is often found in trees and on rocky slopes. It is said to be aggressive and will attack with little provocation.

The reported bites have been inflicted by small (3-foot) individuals. They caused violent headaches with vomiting; both vision and breathing were affected. In one case the victim hemorrhaged from the gums and had local pain, discoloration, and swelling that persisted for several days. The bite of a large snake might be lethal.

No specific antivenin is available for this group of snakes, but "Tiger Snake" antivenin (Commonwealth) is recommended for use.

ELAPIDAE: Genus *Micropechis* Boulenger, 1896.

Pacific coral snakes.

Two species are currently recognized. One of these occurs on New Guinea and some of its offshore islands; the other, *Micropechis elapoides* (Boulenger), with a distinct banded pattern, is found on Florida, Guadacanal, Malaita and Ysabel islands in the Solomons. There are few reports of bites from either of these species. However, they attain lengths of 5 feet and are considered dangerous.

Definition: Head fairly distinct from neck; snout pointed. Body moderately stout, cylindrical; tail short.

Eyes very small; pupils round.

Head scales: The usual 9 on the crown; rostral broad and obtusely pointed. Laterally, nasal in contact with preocular.

Body scales: Dorsals smooth in 15–17 rows throughout body. Ventrals 178–223; anal plate entire or divided; subcaudals generally paired (a few occasionally entire) 35–55.

Maxillary teeth: Two moderately large fangs followed, after an interspace, by 3 small teeth.

Ikaheka Snake. *Micropechis ikaheka* (Lesson).

Identification: Adults average between 3 and 4 feet in length; occasional individuals attain lengths of 5 feet.

Body coloration made up of yellow (or tan) and black (or brown) scales. Black scales roughly arranged in irregular crossbands but each is edged with yellow—sometimes to the extent that the pattern is lost. In specimens from eastern New Guinea the pattern may be lost on the anterior one-third of the body which is brown, but the crossbands persist posteriorly. Belly color yellow with some scutes edged with black.

Dorsals smooth and glossy, in 15 rows at midbody. Ventrals 178–223; 37–55 subcaudals.

Distribution: New Guinea and nearby islands: Aru, Batanta, Mefoor, Mios Num, Misool, Jobi, Mansinam, and Valise.

Remarks: This appears to be the only kind of small-eyed burrowing snake in the Australian-New Guinea region that grows to a size large enough to be a possible hazard. Little seems to be known of its habits; it is apparently a nocturnal or a burrowing species that is seldom seen out during the day. However, at least one death has been reported from its bite. "Tiger Snake" antivenin (Commonwealth Serum Laboratories of Autralia) has been recommended (E. Worrell) for treatment of envenomation from this snake.

ELAPIDAE: Genus *Notechis* Boulenger, 1896.

Australian tiger snake.

A single species, *N. scutatus*, is currently recognized; it has several geographic races and is found in southern Australia and some of the offshore islands. It possesses one of the most toxic venoms known in snakes.

Definition: Head relatively broad, flattened, and moderately distinct from the neck; a distinct canthus rostralis. Body relatively stout, flattened dorsoventrally; tail rather short.

FIGURE 103.—A Black Tiger Snake, *Notechis*. Although only one species of tiger snake is recognized by most workers, Eric Worrell believes these black forms to be a distinct species (*Notechis ater*). Photo by Eric Worrell.

Eyes moderate in size; pupils round.

Head scales: The usual 9 on the crown; frontal wide and shield-shaped. Laterally, nasal in contact with preocular.

Body scales: Dorsals smooth in 17–20 oblique rows at midbody; fewer posteriorly. Ventrals 160–184; anal plate entire; subcaudals single, 43–59.

Maxillary teeth: Two rather long fangs followed, after an interspace, by 3–5 small teeth.

Australian Tiger Snake, *Notechis scutatus* (Peters).

Identification: Adult snakes are 4 to 5 feet long in most parts of the range, but they may exceed 6 feet in Victoria and Tasmania. A record length of 8 feet was reported for a specimen from Chappell Island.

FIGURE 104.—Australian Tiger Snake, *Notechis scutatus*. The most dangerous snake of southern Australia. Photo by Isabelle Hunt Conant. (See also fig. 103)

Ground color varies from yellowish, greenish-gray, orange, and brown to black. The most common pattern is a creamy-yellow ground color banded with gray. Most individuals show a large number of narrow dark bands but those with dark ground colors may be almost unicolor.

Dorsal scales with pointed tips.

Distribution: Tasmania and southern Australia from the border of Queensland to the coastal areas of South Australia. This species inhabits wet areas with rocks and brush.

Remarks: The tiger snake is the most dangerous snake of southern Australia. It is active at night and not aggressive until molested. The greatest danger appears to be from stepping on the snake in the dark. Often there are few local effects from the bite, but the systemic effects are swift and grave.

A specific antivenin (Tiger Snake) is produced by the Commonwealth Serum Laboratories of Australia.

ELAPIDAE: Genus *Ogmodon* Peters, 1865.

Fiji snake.

A single species, *O. vitianus* Peters, is known from Viti Levu and perhaps from other islands of the Fiji group. It is a small burrowing snake; reported lengths are under 20 inches. It is not believed to be a dangerous species.

Definition: Head small and not distinct from neck; no canthus rostralis; snout pointed. Body cylindrical and moderately slender; tail short.

Eyes small; pupils round.

Head scales: The usual 9 on the crown; internasals very small, prefrontals very large and in contact with eye. Laterally, nasal fused to first upper labial; small

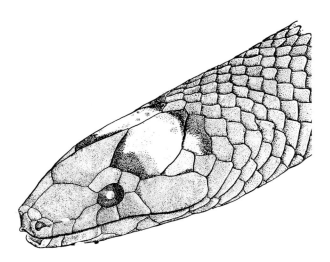

FIGURE 105.—Head Scales of Fiji Snake, *Ogmodon vitianus.* The top of the third upper labial is often separated as a preocular. Drawing by Samuel B. McDowell.

preocular elongate, not in contact with nasal, commonly fused with third upper labial.

Body scales: Dorsals smooth, in 17 rows throughout body. Ventrals 139–152; anal plate divided; subcaudals paired, 27–38.

Maxillary teeth: Two small fangs with external grooves followed, without an interspace, by 5–6 grooved teeth that gradually decrease in length toward the rear.

ELAPIDAE: Genus *Oxyuranus* Kinghorn, 1923.

Taipan.

A single species, *O. scutellatus,* is recognized; it is found in northern Australia and southeastern New Guinea. The taipan reportedly reaches a length of 11 feet. With its long fangs and large supply of very toxic venom, it is considered to be one of the most dangerous snakes living today.

Definition: Head elongate and narrow but distinct

FIGURE 106.—Taipan, *Oxyuranus scutellatus.* The great size and toxic venom make this snake, potentially, one of the most dangerous snakes in the world. Photo by Eric Worrell.

from neck; a distinct canthus rostralis. Body elongate and cylindrical; a moderately long tapering tail.

Eyes large; pupils round.

Head scales: The usual 9 on the crown. Laterally, nasal in contact with preocular.

Body scales: Dorsals with low but distinct keels, in 21–23 rows at midbody, reduced to 17 posteriorly. Ventrals 230–247; anal plate entire; subcaudals 50–72, all paired.

Maxillary teeth: Two long recurved fangs followed, after an interspace, by 2–3 small teeth.

Taipan, *Oxyuranus scutellatus* (Peters).

Identification: Adult taipans average 6 to 7 feet in length; a record specimen was 11 feet long.

Body color is coppery or dark brown in Australian specimens, grayish-black with a reddish-orange stripe along the rear part of the body in New Guinea types.

The skin between the scales is white. Belly is yellow, and speckled with orange in Australian snakes.

The scales are as described in the generic definition.

Distribution: The taipan inhabits grasslands and savannah areas in northern Australia, Melville Island, and coastal New Guinea from the Fly River eastward to the vicinity of Port Moresby. It appears to be most abundant around rocks and boulders, where it lives in rodent burrows.

Remarks: The taipan is active during daylight hours and also on hot nights. It will usually attempt to escape if disturbed, but may become a dangerous adversary if seriously threatened. When provoked it flattens its head, compresses the neck vertically, and expands the body so that the white skin shows between the scales. Adopting a defensive attitude of loose loops, it arches part of the body off the ground and waves its tail. It then attacks so swiftly and suddenly that the victim may be bitten several times before he can defend himself or escape.

This large snake has fangs that are very long for an elapid (over ½ inch in a 7 foot individual) and its venom is one of the most toxic known. Few people survived its bite before a special antivenin was available.

A specific antivenin ("Taipan") is now produced by the Commonwealth Serum Laboratories of Australia.

ELAPIDAE: Genus *Parapistocalamus* Roux, 1934.

Hediger's snake.

A single species, *P. hedigeri* Roux, is known from Bougainville Island, Solomons group. It is a small burrowing snake; the largest known specimen is about 20 inches in length. It is not believed to be a dangerous species.

Definition: Head small and not distinct from neck; no canthus rostralis; snout conspicuously blunted. Body cylindrical and moderately slender; tail short.

Eyes very small; pupils round.

Head scales: The usual 9 on the crown; frontal and prefrontals very broad; rostral broad. Laterally, preocular present or fused with prefrontal; if present preocular in contact with nasal or separated from it by prefrontal.

Body scales: Dorsals smooth, in 15 rows throughout body. Ventrals 159–169; anal plate divided or entire; subcaudals paired, 32–35.

Maxillary teeth: Two fangs of moderate size; no other teeth on maxillary bone.

ELAPIDAE: Genus *Pseudechis* Wagler, 1830.

Australian black snakes and mulga snakes.

Four species are recognized. Three are Australian but one of these (*P. australis*) also occurs in southeastern New Guinea. Another species, *P. papuanus*, is found only in southeastern New Guinea and some of the offshore islands. Both of these species are dangerous.

Definition: Head rather elongate, only slightly distinct from neck; a distinct canthus rostralis. Body depressed and moderately slender; tail moderate in length.

Eyes moderate in size; pupils round.

Head scales: The usual 9 on the crown. Laterally, nasal in contact with preocular.

Body scales: Dorsals smooth and glossy, in 17–21 rows at midbody, reduced to 17 posteriorly. Ventrals 180–230; anal plate divided; anterior subcaudals usually entire, posterior ones paired, 48–70.

Maxillary teeth: Two short fangs followed, after an interspace, by 3–6 small teeth.

Australian Mulga Snake, *Pseudechis australis* (Gray).

Identification: Adult snakes usually measure 5 to 6 feet; a record specimen was "over 9 feet in length."

Body color copper brown. Usually each scale has a red or orange tip and a lighter center, giving a reticulated pattern. Belly cream or yellowish with faint oranges blotches.

Dorsal scales in 17 rows throughout body. There are 180–220 ventrals; 50–70 subcaudals, of which about the first 30 are entire, the remainder paired. (The two known specimens from New Guinea have all entire.)

FIGURE 107.—Australian Mulga Snake, *Pseudechis australis.* Photo by Eric Worrell.

Distribution: This snake is an inhabitant of the dry areas in the northern half of Australia, southern New Guinea, and Melville Island.

Remarks: This large brown snake is often mistaken for the taipan; however, its perfectly smooth scales and fewer ventrals distinguish it from the latter.

The mulga snake is large and relatively aggressive, and will defend itself if held or cornered, flattening the body and neck and striking repeatedly. It will hold on when it bites and chews hard, thus injecting more venom. However, it does not attack unless provoked and its venom rarely causes death.

No specific antivenin is produced but Taipan, Tiger Snake or Papuan Black Snake antivenins are used in treatment; they are all produced by the Commonwealth Serum Laboratories of Australia.

Papuan Black Snake, *Pseudechis papuanus* Peters and Doria.

Identification: Adults are 5 to 7 feet in length.

Body color black or brown, both above and below. Chin whitish. There is no distinct color pattern.

Dorsal scales in 19–21 rows at midbody, 17 rows posteriorly. Ventrals 216–226; subcaudals 49–58, of which the first 21–38 are single, the posterior ones paired.

Distribution: Found only in southeastern New Guinea, Frederick-Hendrik Island, and Yule Island.

Remarks: This snake is closely related to the Australian mulga snake. It is active during the day. Although little is known of its habits, it has a more toxic venom than its relatives.

A specific antivenin ("Papuan Black Snake") is produced by the Commonwealth Serum Laboratories of Australia.

Red-bellied Black Snake. *Pseudechis porphyriacus* (Shaw).

Identification: The average adult size is 5 to 6 feet; record length about 7 feet.

FIGURE 108.—Red-bellied Black Snake, *Pseudechis porphyriacus.* Probably the best known of the poisonous snakes of Australia. Photo by Eric Worrell.

Body color a uniform glossy purplish-black above, and red, pink, or bright orange below.

Dorsal scales in 17 rows throughout body. Ventrals 180–210; 48–66 subcaudals, of which the first 5–20 are usually single, the remainder paired.

Distribution: Swamps, coastal areas, and forested regions of eastern and southern Australia (Queensland, New South Wales, Victoria, and South Australia). It is a good swimmer and is often seen crossing rivers and bays.

Remarks: This is one of the most common and best known of the venomous snakes of Australia. It is active during the day. It is shy and will avoid human contact if given the opportunity. However, it will defend itself with a number of feinted strikes if cornered. Although it bites only under considerable provocation, more bites are recorded for this snake than for any other Australian snake. Less than one percent of the bites are fatal, however, and it is not generally considered a deadly snake.

"Tiger Snake" antivenin, produced by the Commonwealth Serum Laboratories of Australia, is used in the treatment for its bite.

ELAPIDAE: Genus *Rhinoplocephalus*, Müller, 1885.

Müller's snake.

A single species, *R. bicolor* Müller, is known from southern Western Australia. It is small, up to about 16 inches in length, and is not believed to be dangerous. Almost nothing is known of its habits.

Definition: Head small and only slightly distinct from neck; snout broad and flattened. Body cylindrical and moderately slender; tail short.

Eyes small; pupils round.

Head scales: Internasals absent, giving 7 instead of the usual 9 scales on the crown; rostral very broad and slightly free from the other scales on the sides. Laterally, nasal in contact with preocular (with the lower preocular when there are two).

Body scales: Dorsals smooth, in 15 rows at midbody; reduced to 13 posteriorly. Ventrals 149–159; anal plate entire; subcaudals single, 28–34.

Maxillary teeth: Two fangs of moderate size followed, after an interspace, by 2–4 small teeth.

ELAPIDAE: Genus *Rhynchoelaps* Jan, 1858.

Desert banded snake.

Two species are currently recognized; both inhabit the dry regions of Australia. Neither attains a length of more than 16 inches; they are not considered dangerous to man.

Definition: Head small, flattened above but not distinct from neck; snout prominent, with obtusely angular edge; canthus rostralis indistinct. Body cylindrical, moderately slender; tail short.

Eyes small; pupils round.

Head scales: The usual 9 on the crown; rostral broad, obtusely angulate posteriorly; frontal long, much

153

broader than supraoculars. Laterally, nasal long, in contact with preocular.

Body scales: Dorsals smooth, in 15 nonoblique rows throughout body. Ventrals 112–126; anal plate divided; subcaudals paired, 15–25.

Maxillary teeth: Two moderately large fangs with external groves followed, after an interspace, by 3 or 4 small teeth.

ELAPIDAE: Genus *Toxicocalamus* Boulenger, 1896.

Elongate snakes.

Two species are known from New Guinea and the nearby Fergusson, Misima, and Woodlark islands. Neither is known to attain a length of as much as 30 inches; they have short fangs and are not considered dangerous.

Definition: Head small and not distinct from neck; no canthus rostralis. Body cylindrical and very slender; tail short, with distinct terminal spine.

Eyes very small; pupils round.

Head scales: The usual 9 on the crown. Laterally there is no preocular, the prefrontal extends downward to supralabials behind nasal.

Body scales: Dorsals smooth, in 15–17 rows throughout body. Ventrals 230–305; anal plate divided or entire; subcaudals paired, 25–51; terminal subcaudal elongated and compressed, with a keel above.

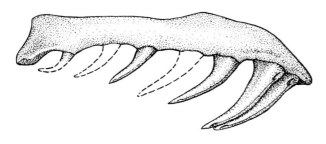

FIGURE 109.—Maxillary Bone of *Toxicocalamus*. Notice fangs and maxillary teeth gradually decreasing in length posteriorly. This is characteristic of a number of elapids in the Pacific Region. Drawing courtesy of Charles M. Bogert.

Maxillary teeth: Two short fangs followed, without an interspace, by 4–5 teeth that gradually decrease in length toward the rear.

ELAPIDAE: Genus *Tropidechis* Günther, 1863.

Rough-scaled snakes.

Two species have been described from eastern coastal Australia, neither of which grows to a length of more than 4 feet. The common species, *T. carinatus* (Krefft), has been reported to inflict a fatal bite.

Definition: Head distinct from neck; a distinct canthus rostralis. Body moderately stout and cylindrical; tail moderately long.

Eyes moderate in size; pupils round.

Head scales: The usual 9 scales on the crown. Laterally, nasal in contact with preocular.

Body scales: Dorsals heavily keeled in 23 rows at midbody. Ventrals 165–216; anal plate entire; subcaudals single or paired, 50–54.

Maxillary teeth: Two large fangs followed, after an interspace, by 4–5 small teeth.

ELAPIDAE: Genus *Ultrocalamus* Sternfeld, 1913.

Short-fanged snake.

A single species is known from New Guinea and the offshore island of Seleo. None of the specimens are as long as 30 inches. The species is not considered dangerous.

Definition: Head small and not distinct from neck; no distinct canthus rostralis. Body cylindrical and quite slender; tail short with blunt tip.

Eyes very small; pupils round.

Head scales: Internasals absent, leaving 7 scales on crown. Laterally there is no preocular, the prefrontal extends downward to supralabials; a small postocular present but parietal extends down to supralabials so that there is no anterior temporal.

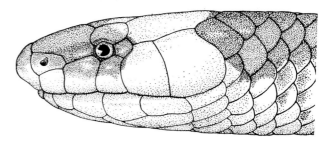

FIGURE 110.—Head Scales of *Ultrocalamus*. Notice absence of preocular, temporals, and internasals characteristic of this genus. Drawing courtesy of Charles M. Bogert.

Body scales: Dorsals smooth, in 13–15 rows throughout body. Ventrals 280–330; anal plate entire; subcaudals paired, 20–54, terminal spine short and flattened.

Maxillary teeth: Two small fangs followed, without an interspace, by 4–6 teeth that gradually decrease in length toward the rear.

ELAPIDAE: Genus *Vermicella* Günther, 1858.

Bandy-bandys.

Five species are recognized. All occur in Australia. None appears to exceed 3 feet in length and they are not considered dangerous.

Definition: Head small and not distinct from neck; no canthus rostralis. Body rather slender and cylindrical; tail very short and obtusely pointed.

Eyes very small; pupils round.

Head scales: The usual 9 scales ordinarily present on crown; the small internasals sometimes fused to prefrontals, giving 7 scales. Laterally, nasal in contact with preocular and first 2 supralabials.

Body scales: Dorsals smooth, in 15 rows at midbody. Ventrals 126–234 (one specimen had 284); anal plate divided; subcaudals paired, 14–30.

Maxillary teeth: Two moderately large fangs followed, after an interspace, by 0–3 small teeth.

REFERENCES
(*See also* General References)

BOGERT, Charles M., and Bessie L. MATALAS. 1945. Results of the Archbold Expeditions. No. 53. A Review of the Elapid Genus *Ultrocalamus* of New Guinea. Am. Mus. Novitates 1244: 8 p., 10 figs.

KINGHORN, J. R. 1956. The Snakes of Australia. 2nd ed. Angus & Robertson Ltd., Sydney. 197 p., illustrated in color.

KINGHORN, J. R., and C. H. KELLAWAY. 1943. The Dangerous Snakes of the South-West Pacific Area. Victorian Railways Printing Works, North Melbourne. 43 p., 11 text figs., 3 pls.

LOVERIDGE, Arthur. 1945. Reptiles of the Pacific World. MacMillan Co., New York. 259 p., 7 pls.

LOVERIDGE, Arthur. 1948. New Guinean Reptiles and Amphibians in the Museum of Comparative Zoology and United States National Museum. Bull. Museum Comp. Zool., 101 (2): 305–430.

ROOIJ, Nelly de. 1917. The Reptiles of the Indo-Australian Archipelago. Vol. II. Ophidia. E. J. Brill Ltd., Leiden. 334 p., 117 text figs.

WERLER, John E., and Hugh L. KEEGAN. 1963. Venomous Snakes of the Pacific Area pp. 217–325, figs. 1–78. *In* H. L. Keegan and W. V. MacFarlane, Venomous and Poisonous Animals and Noxious Plants of the Pacific Region. MacMillan Co., New York.

WILLIAMS, Ernest E., and Fred PARKER. 1964. The Snake Genus *Parapistocalamus* on Bougainville, Solomon Islands (Serpentes, Elapidae). Senckenbergiana-Biol., 45 (3/5): 543–552, figs. 1–5.

WORRELL, Eric. 1963. Dangerous Snakes of Australia and New Guinea. 5th ed. Angus & Robertson Ltd., Sydney. 68 p., illustrated.

WORRELL, Eric. 1963. Reptiles of Australia: Crocodiles, Turtles, Tortoises, Lizards, Snakes.
Angus & Robertson Ltd., Sydney. 207 p., 64 pls. (some in color).

Chapter VIII

DISTRIBUTION AND IDENTIFICATION
of
POISONOUS SEA SNAKES

TABLE OF CONTENTS

TABLE 15.—DISTRIBUTION OF SEA SNAKES

HYDROPHIDAE	Persian Gulf	Arabian Sea	Bay of Bengal	Andaman Sea	Gulf of Siam	S. China Sea	E. China & Yellow Sea	Timor Sea	Sulu Sea	Java & Flores Sea	Banda Sea	Molucca Sea	Celebes Sea	Arafura Sea	Gulf of Carpentaria	Philippine Sea	Coral Sea	Tasman Sea	Southwest Pacific	Central Pacific	Remarks
Acalyptophis peronii		X	X		X	R		X						X	X		X				
Aipysurus apraefrontalis								X			R			X	X						Ashmore Reefs only
A. duboisii					X	X		X		X	R			X	X						
A. eydouxi			X	X	X	X		X		X			R	X	X						Ashmore Reefs only
A. foliosquama								X		R	R										
A. fuscus								X			R			?							
A. laevis		X	X	X	X	X		X		X	X			X	X		X				
Astrotia stokesii		X	X	X	X	X		X	X	X	X	X	X	X	X		X				Also W. Indian Ocean to Seychelles
Emydocephalus annulatus						X		X		X		X		X	X		X				
Enhydrina schistosa	?	X	X	X	X	X	X	X		X	X	X	X	X	X		X				
Ephalophis greyi								X						X	X						Northwest Coast of Australia
Hydrelaps darwiniensis								X						X	X						
Hydrophis belcheri								X		X		X	X	X	X	X	X		X		
H. bituberculatus			X																		Coasts of Ceylon only
H. brookei				X	X	X			X	X				X	X						
H. caerulescens		X	R	X	X	X				X				X	X						
H. cyanocinctus	X	X	X	X	X	X	X		X	X		X		X	X						
H. elegans								X		X				X	X		X				
H. fasciatus		R	X	X	X	X			X	X				X	X		X	X			Also Sea of Japan
H. kingii								X						X	X		X				
H. klossi			R	X	X	X															
H. lapemoides	X	X	X		?																
H. major		X						X						X	X		X				
H. mamillaris	X	X	X	?	?																

(Continued on page 160)

INTRODUCTION

The sea snakes comprise a group of some 50 species all of which have strongly flattened oar-like tails used as sculls. In addition most species have nostrils opening on the top of the head, a body that is flattened from side to side, and very small ventral scutes that may be difficult to distinguish from the adjoining scales. The scales of several kinds of sea snakes are juxtaposed rather than overlapping as in most land snakes. The only snakes likely to be confused with sea snakes are the elephant-trunk snakes (*Acrochordus*) and the river snakes (*Enhydris* and others); these have round or slightly flattened tails, but young elephant-trunk snakes have tails as paddle-shaped as those of some sea snakes. However, all sea snakes have enlarged crown shields and the elephant-trunk snakes have only small juxtaposed scales. Eels are frequently confused with sea snakes; however, no sea snake has fins or gill openings, and none has a smooth skin without scales.

Sea snakes are reptiles essentially of south Asian and Australian coastal waters with a few species found well out into Oceania (Society and Gilbert islands). One species, the pelagic sea snake (*Pelamis*), occurs far out into the open ocean ranging across the Pacific to the western coasts of Central and South America and south to New Zealand and the Cape of Good Hope. No sea snake is found in the Atlantic, although the pelagic sea snake may eventually find its way through the Panama Canal and become established in the Caribbean. The greatest numbers of both species and individuals are found in warm shallow waters without surf or strong currents. Mouths of rivers, bays, and mangrove swamps are especially favored. Many species of sea snakes enter brackish or fresh water occasionally; two species are restricted to lakes.

The biology of sea snakes is poorly known. There is general opinion that they can remain submerged long periods—perhaps a few hours

depending upon temperature, degree of activity, and other factors. The depths to which they can dive are also unknown. An observer in the Philippines saw the snakes swim down out of sight in very clear water. Types of bottom dwelling fish found in stomachs indicate the snakes dive at least 20 to 30 feet to capture food. They are often seen at the surface in calm weather, and some species aggregate there in vast numbers. The reasons for this behavior are unknown, but they may be related to breeding.

There are reports of both diurnal and nocturnal activity. In the Arabian Sea, some species range 10 to 20 miles off shore during the calm winter months but tend to seek coastal mangrove swamps during the monsoon storms. Their young are born in these swamps. Sea snakes feed largely upon fish. Eels are a favorite food of several species. At least a few species eat prawns and one species feeds on fish eggs.

Sea snakes are generally mild tempered reptiles, although both individual and species variation exists with respect to this trait. In open water they either seek to escape or remain almost indifferent to swimmers. Stranded on beaches, some species are almost totally helpless. Others crawl with varying degrees of facility. None can strike on land but most can turn to make an awkward snapping bite. Bites are usually seen when the snakes are slapped, kicked, or trodden upon in shallow water or when they are removed from nets, traps, or other fishing gear.

Some kinds of sea snakes are extensively used for human food in China, Japan, and parts of Polynesia.

While some sea snake species can be identified readily by the amateur, many are puzzling even to experienced herpetologists. Color and pattern are extremely deceptive in this family. There are close similarities between remotely related species and marked differences between young and adults of the same species as well as a good deal of variation among adults of the same species.

KEY TO GENERA

1. A. Ventrals at midbody large, half to one third the width of the belly_____ 2
   B. Ventrals at midbody small or not differentiated_____ 5

TABLE 15.—DISTRIBUTION OF SEA SNAKES (continued)

	Persian Gulf	Arabian Sea	Bay of Bengal	Andaman Sea	Gulf of Siam	S. China Sea	E. China & Yellow Sea	Timor Sea	Sulu Sea	Java & Flores Sea	Banda Sea	Molucca Sea	Celebes Sea	Arafura Sea	Gulf of Carpentaria	Philippine Sea	Coral Sea	Tasman Sea	Southwest Pacific	Central Pacific	Remarks
HYDROPHIDAE (continued)																					
Hydrophis melanosoma				?		X				X			?	X							
H. mertoni														X							
H. nigrocinctus			X	?																	
H. obscurus			X																		Largely confined to brackish water
H. ornatus	R	R	X	X	X	X	X				?	?	?	X	X	?	X	R	X	X	
H. parviceps						X	X														Coast of S. Vietnam only
H. semperi																					(Lake Taal, Luzon, Philippines only)
H. spiralis		X	X	X	?	X	X		R	X	?	?	R								
H. stricticollis		X	X																		
H. torquatus			X	X	X	X				X											
Kerilia jerdonii			X	X	X	X															
Kolpophis annandalei				?		X				R											
Lapemis curtus	X	X	R	?	X	X	X		X	X	X	?	?	R	X	X	R				
L. hardwickii		R	R	?	X	X	X	?	X	X	X	X	X	X	R	X	X	R	X	R	
Laticauda colubrina		R	R	?	X	X	X	X	X	X	X	X	?	R	R	X	R	X	X	R	Pacific coast, Nicaragua (unconfirmed)
L. crockeri																					(Lake Tungano, Rennell Is., Solomons only)
L. laticaudata			R	R	?	X	X	X	X	X	X	X	X	X	X	X	X	X	X		
L. semifasciata						X	X		X	X		X			X	X	X	R	X		
Microcephalophis cantoris		X	X	?																	
M. gracilis	X	X	X	X	X	X	X		X	X	X	X	X	X	X	X	X	X	X	X	
Pelamis platurus *	R	X	X	X	X	X	X	R	X	X	X	X	X	R	X	X	X	R	X	R	
Praescutata viperina	X	X	X	X	X	X	X		X	X							?		X	X	
Thalassophis anomalus					X	X			X	X									X	X	

The symbol X indicates widespread distribution of the species within the area, however it is emphasized that most sea snakes are restricted to coastal water. The symbol R indicates an isolated record or a very few records possibly based on strays or waifs. The symbol ? indicates suspected presence of the species within the area.

*Also western & southern Indian Ocean; Sea of Japan north to Possiet Bay, Kamchatka; eastern Pacific along American coasts from Gulf of California to Ecuador.

KEY TO GENERA (continued)

2. A. Nostrils lateral; nasal shields separated by inter-
        nasals (fig. 111A), 4 species; widely distributed
        from Bay of Bengal to central Pacific_____ *Laticauda*
  B. Nostrils dorsal; nasal shields in contact with each
        other (see fig. 111B)_____ 3
3. A. Tail distinctly paddle-shaped; head shields entire
        or broken up_____ 4
  B. Tail but slightly paddle-shaped, almost round;
        head shields entire (single species, small and
        rare)_____ *Ephalophis*
4. A. Snout smoothly rounded; fangs followed by sev-
        eral small teeth on maxillary bone (Six or 7
        species of moderate to large size; found from
        Gulf of Siam to Coral Sea but mostly in Aus-
        tralian and New Guinea waters. Larger species
        potentially dangerous but nothing known of
        venoms.)_____ *Aipysurus*
  B. Snout has blunt spine directed forward; fangs very
        small, no other maxillary teeth (single species
        ranging from Ryukyus to Australia; inefficient
        biter; probably not dangerous; venom un-
        known)_____ *Emydocephalus*
5. A. Ventrals distinct on at least the posterior half of
        the body, not normally split, usually a little
        larger than adjacent scales_____ 6
  B. Ventrals, except quite anteriorly, divided by a fis-
        sure or very small and not well differentiated
        from adjacent scales_____ 13
6. A. Mental shield elongated and concealed in cleft (fig.
        111C); ventrals often not well differentiated on

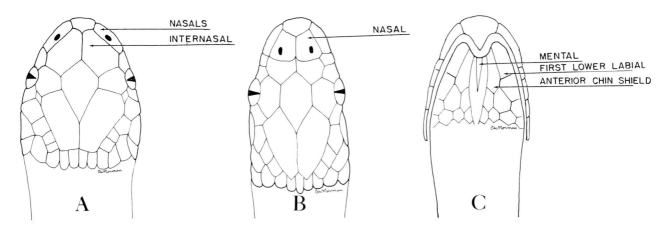

FIGURE 111.—A. Top of head of sea krait (*Laticauda*) showing separation of nasals by internasals; B. Top of head of sea snake showing nasals in contact with each other; C. Lower jaw of beaked sea snake showing elongate mental in cleft between chin shields.

KEY TO GENERA (continued)

anterior half of body. Widely distributed, abundant species—most dangerous of sea snakes_____ *Enhydrina*

    B. Mental shield normal; ventrals well differentiated the entire length of the body_____ 7

7. A. Head shields entire; nasal shields in contact with each other_____ 8

    B. Head shields more or less divided_____ 11

8. A. Preocular shield absent; tail not strongly paddle-shaped; ventrals almost one fourth width of belly. Single, small, brightly colored species; not believed dangerous_____ *Hydrelaps*

    B. Preocular shield present; tail distinctly paddle-like; ventrals smaller, at least on posterior half of belly_____ 9

9. A. Scale rows around middle of body 19–23. Single species found in Indo-Malaysian waters and locally plentiful. Bites have been reported, but are not serious_____ *Kerilia*

    B. Scale rows around middle of body 25 or more_____ 10

10. A. Ventrals decidedly large on anterior quarter of body, much smaller posteriorly. Single species, widely distributed, venom unknown_____ *Praescutata*

    B. Ventrals more or less same size entire length of the body. Twenty-two species widely distributed; in several species bite produces death_____ *Hydrophis*

11. A. Nasals contact each other. Scales around eye with spiny projections; body scales with pointed keels. Single rare species; large and considered dangerous_____ *Acalyptophis*

    B. Nasals separated by internasals; no spines on head; body scales without pointed keels_____ 12

12. A. Dorsal scales in 31–35 regular rows at midbody. Single species_____ *Thalassophis*

    B. Dorsal scales very small, in 75–93 irregular rows at midbody. Single species_____ *Kolpophis*

13. A. Head very small, neck long and slender. Two species, one widely distributed_____ *Microcephalophis*

    B. Body form not as above_____ 14

14. A. Dorsal scales overlapping (imbricate). Single large species_____ *Astrotia*

    B. Dorsal scales juxtaposed_____ 15

15. A. Head elongated, flat; all body scales quadrangular, generally equal in size. Single species with widest distribution of any sea snake_____ *Pelamis*

B. Head short, chunky; 3 or 4 rows of larger scales on flanks; anterior ventrals often enlarged. Two species range from Persian Gulf to Japan and south to Australia. Considered dangerous_____ *Lapemis*

SPECIES DESCRIPTIONS

Yellow-lipped Sea Krait, *Laticauda colubrina* (Schneider).

Identification: Species of this genus are less flattened and more like conventional land snakes than are other members of the family. They can be readily identified by the combination of flattened tail, enlarged ventral scutes, and laterally placed nostrils. In this species the pattern consists of black or dark brown bands encircling body and separated by interspaces of pale blue or blue gray ground color; these are about as wide as the bands; snout and upper lip yellow; dark stripes through eye and on lower lip; belly yellow.

FIGURE 112.—Yellow-lipped Sea Krait, *Laticauda colubrina*. Photo by Robert E. Kuntz.

Maximum length about 4½ feet, average 3 to 3½ feet. Females are larger than males.

Remarks: One of the few sea snakes that regularly leaves water to climb onto rocks and pilings. Terrestrial activity usually takes place at night. Eggs are deposited in caves and crevices. Very mild disposition—no report of bite in man although the snakes are freely handled by many natives. Venom of fairly high toxicity but very small in quantity.

Beaked Sea Snake, *Enhydrina schistosa* (Daudin).

Identification: The distinctive feature of this sea snake is the form of the lower jaw. The shield at the tip of the chin (the mental) which is comparatively wide and large in most snakes is, in *Enhydrina*, reduced to a splinterlike shield buried in a cleft between the first pair of lower labials (fig. 111C). This gives greater flexibility to the lower jaw and widens the gape thus permitting the snake to seize and swallow large prey. The down

curved tip of the rostral is unusually prominent in this snake giving it a characteristic beaked profile. Head shields large, symmetrical; head rather small, very little wider than neck; body moderately stout, strongly compressed; skin especially on neck rather loose; scales keeled; ventrals poorly differentiated, often indistinguishable on anterior part of body.

Adults uniformly dull olive green above or pale greenish gray with dark crossbands that tend to fuse anteriorly; cream to dirty white on sides and belly; head greenish above without marking; tail usually mottled with black. New born young are milk white with crossbands that almost encircle the body; top of head dark olive, tail black.

Average adult length 3 to 4 feet with females appreciably larger than males; maximum a little under 5 feet.

Remarks: A shallow water snake found over both mud and sand bottom and often very plentiful at the mouths of rivers. In great deltas such as those of the Ganges and Indus, *Enhydrina* has been found in channels many miles from the open sea. It has not been reported to leave the water voluntarily and is very

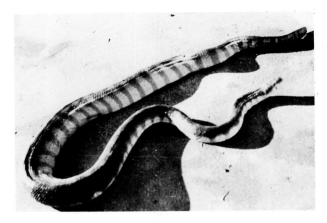

FIGURE 113.—Beaked Sea Snake, *Enhydrina schistosa*. Photo by Sherman A. Minton.

awkward although not completely helpless on land. In Indian waters, young are born from March through July. The average brood numbers 4 to 9.

The venom of the beaked sea snake is the most toxic of the better known snake venoms, the lethal dose for experimental mammals being 50 to 125 micrograms/

kilo of body weight. Since the fatal dose for an adult man is estimated to be about 1.5 mg. and 10 to 15 mg. can be obtained from a snake of average size, this is clearly a potentially dangerous species, and it does appear to be responsible for more serious and fatal bites than all other sea snakes combined. This may be ascribed partly to its very toxic venom, and partly to its abundance near bathing beaches and fishing villages. It is ordinarily an inoffensive reptile but will bite if restrained. An antivenin against the venom of this sea snake is produced by Commonwealth Serum Laboratories, Victoria, Australia. It is reported to be effective against venoms of the larger species of *Hydrophis* also.

Olive-brown Sea Snake, *Aipysurus laevis* Lacépède.

Identification: In sea snakes of this genus the nostrils are dorsal in position and the shields on the top of the head are small but regular in arrangement. The ventrals are well developed extending at least one-third the width of the body. *Aipysurus laevis* is a very heavy snake, often as thick as a man's arm; the body is slightly flattened vertically. The head is large and a little wider than the neck; the end of the tail is usually ragged.

Adults are uniformly olive brown or may have a row of dark spots on the flanks and belly.

Maximum length about 6 feet; average 4½ to 5 feet.

Remarks: Clumsy on land, it apparently does not leave water voluntarily although it is often found stranded on beaches.

Nothing is known of the venom nor are there reliable reports of bites.

Stokes's Sea Snake, *Astrotia stokesii* (Gray).

Identification: Like the olive-brown sea snake (*Aipy-*

surus laevis) in being of massive build with a large head; differs in having ventrals that are fragmented and not well differentiated and larger head shields. The body scales are large, keeled, and strongly imbricate.

Color light brown, yellowish or orange above with broad black rings or bars and spots; belly paler; head olive to yellowish.

Average adult length 4½ to 5 feet; maximum about 6 feet; large specimens are about 10 inches in girth.

Remarks: Although generally an uncommon snake there is a report of a vast aggregation sighted in Malacca Strait on the 4th of May. The snakes were disposed in a line about 10 feet wide and some 60 miles long. This appears to be a snake of moderately deep open water and is not often taken by native fishermen. There are no reported bites by this species and its venom has never been studied.

The sea snake genus *Hydrophis* is a puzzling one from the standpoint of classification, and exact identification of most of the 25 or so species requires expert opinion. These snakes have the characteristic sea snake features of laterally compressed body and tail, nostrils located on the top of the head in nasal shields that are in contact with each other, small eyes with round pupils and absence of the loreal shield. The ventrals are small but generally larger than the adjoining scales and form a distinct series that in large adults of many species is transformed into a keel or ridge. The four species described here are common, widely distributed, and show something of the range of variation encountered.

FIGURE 115.—Yellow Sea Snake, *Hydrophis spiralis.* Photo by Sherman A. Minton (Preserved specimen).

Yellow Sea Snake, *Hydrophis spiralis* (Shaw).

Identification: Head of moderate size and distinct from the neck which is not particularly slender or elongated; body moderately slender, not strongly compressed. Head shields large and symmetrical; the tip of the rostral shows a slight downward prolongation that fits into a notch in the tip of the lower jaw; usually one anterior temporal; body scales smooth or weakly keeled. There is an increase of no more than 8 scale rows be-

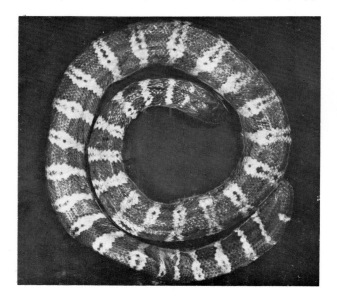

FIGURE 114. Stokes's Sea Snake, *Astrotia stokesii.* Photo by Edward H. Taylor (Preserved specimen).

tween a count made on the neck and one made at the middle of the body.

Color golden yellow to yellowish green shading to pinkish white below; body encircled by black rings that are widest along the vertebral midline and narrow on the flanks, always much narrower than the interspaces separating them; head uniformly yellow in the adult, dark with a yellow horseshoe shaped mark on the crown in the young.

This is apparently the longest of the sea snakes, although *Aipysurus* and *Astrotia* exceed it in bulk. Adult yellow sea snakes frequently reach a length of 5½ to 6 feet, and a record length of 9 feet is reported.

Remarks: Very little information is on record concerning the habits and biology of this sea snake. It seems to frequent deep water and often basks at the surface.

Venom yields from this snake are surprisingly small (3 to 10 mg.) and toxicity lower than for most sea snake venoms, nevertheless several fatalities are on record from the bite of this species.

Annulated Sea Snake, *Hydrophis cyanocinctus* Daudin.

Identification: Head smaller, neck longer and more slender and body more compressed than in the yellow sea snake. Head scales similar to those of the yellow sea snake except that there are usually 2 anterior temporals. Body scales with central keel or row of tubercles. Increase of more than 8 (usually 10 to 16) scale rows between count at neck and count at midbody.

FIGURE 116.—Head Scales of *Hydrophis cyanocinctus*. (See also plate VI, fig. 4.) Redrawn from Maki, 1931.

Color dirty white, pale greenish, yellow or olive with blackish crossbands that may or may not encircle the body, are widest along the vertebral midline and are as wide as, or wider than, the interspaces between them. Head in adult olive, reddish, or dull yellow; in young blackish with the yellow horseshoe mark seen in some other species.

The adult length averages 4½ to 5½ feet with record specimens of about 6½ feet.

Remarks: This snake frequents mangrove swamps but has been collected 12 to 20 miles offshore during winter. Although it has not been seen to leave the water voluntarily, it crawls fairly well and can lift its head well free of the ground. It often bites if restrained.

Venom yields reported from this snake are approximately double those reported for the yellow sea snake, and the toxicity is somewhat higher. Data from Malaya indicate *H. cyanocinctus* causes more deaths than any sea snake species except the beaked sea snake.

Reef Sea Snake, *Hydrophis ornatus* (Gray).

Identification: A large headed, stout bodied sea snake; body scales small, juxtaposed, with a central tubercle that is more strongly developed in the male; increase of 12 to 20 scale rows between count at neck and count at midbody. The combination of regular head shields with nasals in contact with each other and small, undivided ventrals of almost uniform size the entire length of the body will usually differentiate this species from other sea snakes of similar body build.

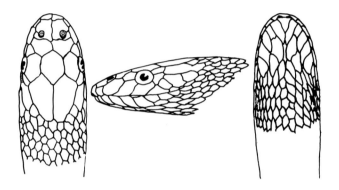

FIGURE 117.—Head Scales of *Hydrophis ornatus*. Note regular head scutes and contact of nasals. Redrawn from Maki, 1931.

The typical form is pale greenish white, olive or yellow with wide dark crossbands or rhomboid spots. The head is olive. The Philippine subspecies is uniformly grayish green above and whitish below. A subspecies with spotted or ocellate markings on the sides occurs in Australian waters.

Average length is 28 to 35 inches; maximum about 45 inches.

Remarks: This sea snake has a very wide range extending from the Persian Gulf to the central Pacific and from the Yellow Sea to Australia. It is plentiful in some localities, e.g. Manila Bay, but very rare and apparently only a straggler in many others. It evidently frequents shallow water, for dozens have been taken in one haul of a beach seine. At least one fatality is ascribed to its bite.

Banded Small-headed Sea Snake, *Hydrophis fasciatus* (Schneider).

Identification: Certain species of the genus *Hydrophis* and the two species of *Microcephalophis* are remarkable for their tiny heads and long slender necks. This peculiar body form is most evident in the adult;

young are not strikingly different from other sea snakes. Recognition of the young and differentiation of the various species of small-headed sea snakes is difficult. In this species the ventrals are distinctly wider than the adjacent scales throughout the length of the body and the total ventral count is high, usually 400 or more.

Thick part of body gray to dirty yellow crossed by dark bands that are widest in the middle of the back, but taper to points laterally; neck dark olive to black with yellow spots or crossbars; head uniformly dark.

The maximum length does not exceed 4 feet; average specimens are about 3 feet.

Remarks: The heavy body gives stability in floating while the small head and long slender neck permit the snake to explore holes and crevices in search of the eels and other elongate fishes that are its food. In swimming free, the head and neck are held straight and almost motionless. This species is reported to be primarily nocturnal in the Philippines.

Small-headed sea snakes are among the least prolific of snakes, females giving birth to only 1 or 2 young in a season.

Although it is difficult to imagine these reptiles biting effectively, there is at least one fatality ascribed to the bite of a small-headed sea snake. Venom yields are minute (less than 1 mg. per snake), but the venom is extremely toxic, being about equal to that of *Enhydrina* (p. 163).

Other widely distributed small-heads of the genus *Hydrophis* include *H. belcheri* of Australian and Pacific seas and *H. brookei*, *H. caerulescens*, and *H. klossi* of Indo-Malaysian waters.

Graceful Small-headed Sea Snake, *Microcephalophis gracilis* (Shaw).

Identification: This snake differs from the banded small-headed sea snake in certain features of the skull

FIGURE 118.—Graceful Small-headed Sea Snake, *Microcephalophis gracilis*. The small end is the head. U.S. Navy photo.

and in the type of ventrals which are distinctly wider than the adjacent scales on the slender part of the body but become smaller and fragmented posteriorly; the ventral count does not exceed 300.

Anterior part of body including head black to dark olive with white or yellow spots or bands; posterior part pale yellow to greenish with darker crossbands or uniform gray above and light laterally and ventrally.

Most adults of this species measure 30 to 35 inches; maximum length is about 42 inches.

Hardwicke's Sea Snake, *Lapemis hardwickii* Gray.

Identification: A rather short, stocky sea snake; head chunky, wider than neck; rostral with 3 stubby downward projections fitting into notches in the chin; ventrals not well differentiated except on neck; irregular rows of enlarged scales low on flanks.

FIGURE 119.—Hardwicke's Sea Snake, *Lapemis hardwickii*. Photo by Edward H. Taylor (Preserved specimen).

Greenish or yellowish above, with series of dark crossbands that are much wider than the light areas between them; paler below; head dark with or without lighter mottling; tail barred with black tip.

The average length of adults is 25 to 30 inches with a maximum of about 35 inches.

Remarks: A very abundant snake in shallow estuaries along coasts of Viet Nam, Malaya, and the Philippines and often taken in fish nets. It is most abundant during the rainy season (July to November).

Despite the very small venom yield (about 2 mg. from an adult snake) several fatal bites are on record; toxicity of the venom is less than that of the beaked sea snake (p. 163).

Pelagic Sea Snake, *Pelamis platurus* (Linnaeus).

Identification: Head elongate, flat, slightly wider than neck; body of moderate build, very strongly compressed laterally; the entire appearance is very eel-like. Head shields large, symmetrical; body scales small, quadrangular; ventrals not larger than adjacent scales.

The commonest color variety is black or dark brown above, dark yellow to brown below with a pale yellow lateral stripe. Another common variety is yellow with a straight-edged brown or black dorsal stripe. In less common forms, the dark stripe may be wavy or broken into transverse bars. The head is usually dark on top and yellow on the sides; the tail is whitish barred or mottled with black.

The average adult length is 25 to 30 inches with a maximum of 44 inches.

FIGURE 120.—Pelagic Sea Snake, *Pelamis platurus*. The most widespread species of sea snake; the one found along the west coast of tropical America. U. S. Navy photo.

Remarks: This is the only truly ocean-going snake; it has repeatedly been seen hundreds of miles from land and has reached many remote Pacific islands including Hawaii. It is, nonetheless, the most plentiful in the comparatively shallow waters over the continental shelves. Although a graceful, rapid swimmer, it seems to spend much time floating at the surface. It is virtually helpless on land. Great schools of these snakes have been seen in the shallow waters along the west coast of tropical America at certain seasons. This is a species that seems to be definitely repelled by fresh or brackish water and does not enter creeks or rivers.

Only minute amounts of venom can be obtained from this species in the laboratory, and the toxicity is about one fourth that of *Enhydrina* (p. 163). Only one human fatality has been ascribed to the bite of *Pelamis;* the report dates back almost a century and the snake may not have been correctly identified.

REFERENCES

BARME, Michel. 1963. Venomous Sea Snakes of Viet Nam and Their Venoms. *In* Venomous and Poisonous Animals and Noxious Plants of the Pacific World (H. L. Keegan and W. V. MacFarlane, eds.). Pergamon Press: Oxford, pp. 373–378, figs. 1–5.

HERRE, Albert. 1942. Notes on Philippine Sea-snakes. Copeia, no. 1, pp. 7–9. 1949. Notes on Philippine Sea Snakes of the Genus *Laticauda. Ibid.*, no. 4, pp. 282–284.

SMITH, Malcolm A. 1926. A Monograph of the Sea-snakes. Taylor and Francis: London, pp. i–xvii + 1–130, figs. 1–35, pls. 1–2.

VOLSØE, Helge. 1939. The Sea Snakes of the Iranian Gulf. *In* Danish Scientific Investigations in Iran (Knud Jessen and Ragnar Sparck, eds.) Copenhagen, Part 1, pp. 9–45.

NOTES

Chapter IX

ANTIVENIN SOURCES[*]

Antivenins are available for use in the treatment of most cases of snake venom poisoning. They may be supplied as the whole serum taken from horses immunized against the venom(s) and packaged as a solution or as a dried product, or they may be supplied in a concentrated or purified form as either a liquid or a lyophilized powder. The ability of these various products to neutralize a specific venom may differ considerably. In general, the more species and genus specific the antivenin the more effective it will be in combating the effects of poisoning. Also, the more concentrated and purified the product the more effective it will be and the less the likelihood to produce anaphylaxis. Before administering any antitoxin the physician or corpsman should consult the brochure accompanying the antivenin for specific instructions.

The following tabulation shows the antivenins available, the country in which they are produced, the producer, the name of the product, the venom(s) used in the preparation of the antivenin, the common name of the snake, comments on the additional venoms that the antivenin may neutralize, and data on the processing of the antivenin. The venom(s) used in the preparation of the antivenins appear singly (monovalent) or in groups (polyvalent).

REFERENCES

RUSSELL, F. E. and LAURITZEN, L. 1966.
  Antivenins. Trans. Royal Soc. Trop. Med.
  Hyg., *60*: 797–810.

---

[*] This list of sources and types of antivenins is accurate within the limits of the latest available information from manufacturers at the time of publication.

ANTIVENIN SOURCES

COUNTRY	PRODUCER	NAME OF PRODUCT	VENOM(S) USED IN PREPARATION	COMMENTS
ALGERIA	Institut Pasteur d'Algerie, Rue Docteur Laveran, Alger	Serum Antiviperin	Cerastes cerastes (African desert horned viper), Vipera lebetina (probably the species referred to in this manual as V. mauritanica, Sahara rock viper)	Refined and concentrated.
ARGENTINA	Instituto Nacional de Microbiologia, Vélez Sarsfield 563, Buenos Aires	Monovalente Anti-Crotálico	Crotalus durissus terrificus (Cascabel)	All antivenins are purified by enzymatic and differential thermocoagulation techniques.
		Bivalente Anti-Botrópico	Bothrops alternatus (Urutu), B. neuweidi (Jararaca pintada)	
		Polivalente	Crotalus durissus terrificus (Cascabel), Bothrops alternatus (Urutu), B. neuweidi (Jararaca pintada)	
		Polivalente Misiones	Crotalus durissus terrificus (Cascabel), Bothrops alternatus (Urutu), B. neuweidi (Jararaca pintada), B. jajaraca (Jararaca), B. jararacussu (Jararacussú)	
AUSTRALIA	Commonwealth Serum Laboratories, Poplar Road, Parkville, Melbourne, Victoria	Brown Snake Antivenin	Demansia textilis (Australian brown snake)	Brown Snake Antivenin neutralizes venoms of other species of Demansia such as the black whip snake. Tiger Snake Antivenin neutralizes the venoms of all important venomous snakes of the Australian and New Guinea region to some extent. Taipan Antivenin neutralizes Australian brown snake and mulga snake venoms, while Papuan Black-snake Antivenin neutralizes
		Death Adder Antivenin	Acanthophis antarcticus (Death adder)	
		Malayan Cobra Antivenin	Naja n. sputatrix (Malayan cobra)	
		Papuan Black Snake Antivenin	Pseudechis papuanus (Papuan blacksnake)	
		Sea Snake Antivenin	Enhydrina schistosa (Beaked sea snake)	

Taipan Antivenin — *Oxyuranus scutellatus* (Taipan)

Tiger Snake Antivenin — *Notechis scutatus* (Tiger snake)

Papua-New Guinea Polyvalent Antivenin — *Oxyuranus scutellatus* (Taipan), *Acanthophis antarcticus* (Death adder), *Notechis scutatus* (Tiger snake), *Pseudechis papuanus* (Papuan blacksnake)

red-bellied blacksnake and mulga snake venoms. Recent animal experiments (Minton, unpublished data) indicate that these antivenins also have considerable ability to neutralize venoms of cobras (*Naja* and *Ophiophagus*) and kraits (*Bungarus*). When used against venoms of heterologous species, the doses of antivenin should be two to four times as large as those recommended for the homologous species. Sea Snake Antivenin may be expected to have some neutralizing effect against venoms of sea snakes other than *Enhydrina schistosa* but clinical data are lacking. All of the Commonwealth Serum Laboratories antivenins are prepared by pepsin digestion and ammonium sulfate precipitation. The products are dialyzed and ultrafiltered to a final concentration of 20% protein.

AUSTRIA————Serotherapeutisches Institute Wien, Triesterstrasse 50, Wien

Acts as distributor for antivenins produced by Behringwerke AG, Marburg-Lahn, Germany.

BRAZIL————Instituto Butantan, Caixa Postal 65, São Paulo

Sôro Anti-Crotálico — *Crotalus* sp.

Sôro Anti-Laquético — *Lachesis mutus* (Bushmaster)

Sôro Anti-Elapídico — *Micrurus* sp.

Sôro Anti-Botrópico — *Bothrops* sp.

Sôro Anti-Ophídico — *Bothrops* ssp. and *Crotalus* sp.

Sôro Anti-Elapídico is reported to be effective in treatment of envenomation by other species of coral snakes. Sôro Anti-Botrópico and Anti-Ophídico are reported to neutralize the venom of the bushmaster (*Lachesis mutus*) in large dosage.

These antivenins are prepared by pepsin digestion and ammonium sulfate precipitation.

Poisonous Snakes of the World

ANTIVENIN SOURCES (Con't.)

COUNTRY (Cont'd)	PRODUCER (Cont'd)	NAME OF PRODUCT (Cont'd)	VENOM(S) USED IN PREPARATION (Cont'd)	COMMENTS (Cont'd)
	Instituto Pinheiros Productos Therapeuticos, Caixa Postal 951, São Paulo	Sôro Anti-Botrópico	*Bothrops alternatus* (Urutu), *B. atrox* (Barba amarilla), *B. jajaraca* (Jararaca), *B. jararacussu* (Jararacussú), *B. cotiara* (Cotiara)	A monovalent *Bothrops jajaraca* (Jararaca) antivenin is also available. Potency period 3 years. Sôro Anti-Botrópico and Anti-Ofídico are reported to neutralize the venom of the Bushmaster (*Lachesis mutus*) in large doses. These antivenins are prepared by pepsin digestion and ammonium sulfate precipitation. The final solution contains 18% protein.
		Sôro Anti-Ophídico	Same as Anti-Botrópico plus *Crotalus durissus terrificus* (Cascabel)	
		Sôro Anti-Crotálico	*Crotalus durissus terrificus* (Cascabel)	
COLOMBIA	Instituto Nacional de Salud, Calle 57, Numero 8-35, Bogotá, D.E.	Suero Anti-Ofídico	*Crotalus durissus* ssp. (Cascabel), *Bothrops* sp.	Concentrated by ammonium sulfate precipitation and lyophilized. (Production discontinued March 1966.)
FRANCE	Institut Pasteur, 36 Rue du Docteur Roux, Paris, 15	Serum Antivenimeux Cobra	*Naja naja* (Asiatic cobra)	Serum Antivenimeux Cobra is reported to neutralize the venom of the Egyptian cobra (*Naja haje*) as well as the Asian species. Serum Antivenimeux Naja is reported to neutralize the venom of other African cobras. *Naja haje*, *N. nigricollis*, and *Bothrops lanceolatus* monovalent antivenins are also available.
		Serum Antivenimeux Naja	*Naja haje* (Egyptian cobra), *N. nigricollis* (Spitting cobra)	
		Serum Antivenimeux Dendraspis	*Dendroaspis viridis* (Western green mamba)	
		Serum Antivenimeux Bungarus	*Bungarus flaviceps* (Red-headed krait)	
		Serum Antivenimeux Ancistrodon	*Agkistrodon rhodostoma* (Malayan pit viper)	
		Serum Antivenimeux Echis-Naja	*Echis carinatus* (Saw-scaled viper), *Naja haje* (Egyptian cobra), *N. nigricollis* (Spitting cobra)	

Serum Antivenimeux Berus-Ammodytes	*Vipera berus* (European viper), *V. ammodytes* (Long-nosed viper), *V. aspis* (Asp viper)	
Serum Antivenimeux Echis	*Echis carinatus* (Saw-scaled viper)	Not concentrated.
Serum Antivenimeux Bitis	*Bitis arietans* (Puff adder), *B. gabonica* (Gaboon viper)	
Serum Antivenimeux Cerastes	*Cerastes cerastes* (African desert horned viper). *C. vipera* (Sahara sand viper)	
Serum Antivenimeux Lebetina-Xanthina	*Vipera lebetina* (Levantine viper), *V. xanthina* (Near East viper), *V.x. palaestinae* (Palestine viper)	
Serum Antivenimeux Bitis-Echis-Naja	*Bitis arietans* (Puff adder), *B. gabonica* (Gaboon viper), *Echis carinatus* (Saw-scaled viper), *Naja nigricollis* (Spitting cobra), *N. haje* (Egyptian cobra)	A 12-13% globulin solution.
Serum Antivenimeux Aspis-Berus	*Vipera aspis* (Asp viper). *V. berus* (European viper)	Supplied as whole serum or 12-13% globulin solution.
GERMANY——Behringwerke AG, Postschliessfach 167, 355 Marburg		
Serum Nordafrika	*Bitis gabonica* (Gaboon viper), *B. arietans* (Puff adder), *Cerastes cerastes* (African desert horned viper), *C. vipera* (Sahara sand viper), *Vipera lebetina*[1] (Levantine viper), *Echis carinatus* (Saw-scaled viper), *Naja nigricollis* (Spitting cobra), *N. haje* (Egyptian cobra)	Concentrated but not lyophilized. Final solution 16% protein. Prepared by pepsin digestion and ammonium sulfate precipitation.
Serum Zentralafrika	*Bitis gabonica* (Gaboon viper), *B. arietans* (Puff adder), *B. nasicornis* (River jack), *Hemachatus haemachatus*	

ANTIVENIN SOURCES (Con't.)

COUNTRY (Cont'd)	PRODUCER (Cont'd)	NAME OF PRODUCT (Cont'd)	VENOM(S) USED IN PREPARATION (Cont'd)	COMMENTS (Cont'd)
			(Ringhals), *Naja nigricollis* (Spitting cobra), *N. haje* (Egyptian cobra)	
		Serum Naher- und Mittlerer Orient	*Cerastes cerastes* (African desert horned viper), *Echis carinatus* (Saw-scaled viper), *Vipera ammodytes* (Long-nosed viper), *V. lebetina*[1] (Levantine viper), *Naja haje* (Egyptian cobra)	
		Serum Europa	*Vipera ammodytes* (Long-nosed viper), *V. lebetina*[1] (Levantine viper)	
		Serum Mittle- und Sudamerika	*Crotalus durissus terrificus* (Cascabel), *Bothrops atrox* (Barba amarilla), *B. jajaraca* (Jararaca)	
		Serum Kobra	*Hemachatus haemachatus* (Ringhals), *Naja haje* (Egyptian cobra), *N. nivea*[1] (Yellow cobra), *N. nigricollis* (Spitting cobra)	
INDIA	Central Research Institute, Kasauli, R.I., Punjab	Polyvalent Antivenin	*Bungarus caeruleus* (Indian krait), *Naja naja* (Indian cobra), *Echis carinatus* (Saw-scaled viper), *Vipera russelii* (Russell's viper)	Concentrated and lyophilized.
	Haffkine Institute, Parel, Bombay 12, Maharashtra	Polyvalent Anti-Snake Venom Serum	Same as in Central Research Institute polyvalent antivenin.	Digested with pepsin. Concentrated and lyophilized.
INDONESIA	Perusahaan Negara Bio Farma, (Pasteur's Institute) 9 Djalan Pasteur, Postbox 47,	ABM Antivenin	*Bungarus fasciatus* (Banded krait), *Naja n. sputatrix* (Malayan cobra), *Agkistrodon*	Not concentrated or lyophilized. Available as monovalent antivenins. Potency period 2 years.

	Bandung	*rhodostoma* (Malayan pit viper)	
IRAN	Institut d'Etat des Serums et Vaccins Razi, Boîte postale 656, Téhéran (Hessarak-Karadj)	Polyvalent Antivenin for Iran	*Naja n. oxiana* (Oxus cobra), *Vipera lebetina* (Levantine viper), *Echis carinatus* (Saw-scaled viper), *Vipera persica* (referred to in this work as *Pseudocerastes persicus*, the Persian horned viper), *Vipera xanthina* (probably *V. x. raddei*, the Elburz Mountain viper), *Agkistrodon halys* (probably *A. h. caucasicus*, Pallas' pit viper), *Vipera ammodytes* (Long-nosed viper), *Vipera cerastes* (probably *Cerastes cerastes vipera*, the Sahara sand viper), *Cerastes cerastes* (African horned desert viper)	Monovalent antivenins against venoms of *Pseudocerastes persicus* (Persian horned viper), *Vipera x. raddei* (Elburz Mountain viper), and Pallas' pit viper (*Agkistrodon halys caucasicus*) are produced in limited quantity, and not for general distribution. All antivenins are prepared by pepsin digestion and ammonium sulfate precipitation. Potency period 3 years.
		Polyvalent Antivenin for Middle East, India, and Pakistan	*Naja naja* (Indian cobra), *Bungarus fasciatus* (Banded krait), *Vipera russelii* (Russell's viper), *Echis carinatus* (Saw-scaled viper), *Vipera lebetina* (Levantine viper)	
ISRAEL	Rogoff Wellcome Research Laboratories, Beilinson Hospital, P.O.B. 85, Petah Tikva	Echis coloratus Antiserum	*Echis coloratus* (Arabian saw-scaled viper)	Whole venom plus resin-bound neurotoxin used as antigen; supplied as whole serum in liquid form.
		Vipera palaestinae Antiserum	*Vipera palaestinae* (Palestine viper)	
ITALY	Instituto Sieroterapico e Vaccinogeno "SCLAVO" Via Fiorentina, Siena	Antiophidio Serum	*Vipera ammodytes* (Long-nosed viper), *V. aspis* (Asp viper), *V. berus* (European viper)	Reported to be effective against venoms of all European vipers. Not concentrated, supplied in dried form.

ANTIVENIN SOURCES (Con't.)

COUNTRY (Cont'd)	PRODUCER (Cont'd)	NAME OF PRODUCT (Cont'd)	VENOM(S) USED IN PREPARATION (Cont'd)	COMMENTS (Cont'd)
JAPAN	Institute for Medical Science, University of Tokyo, Shiba Shirokane-daimachi Minato-ku, Tokyo	Mamushi-venom Antivenin	*Agkistrodon halys blomhoffii* (Mamushi)	All antivenins concentrated and lyophilized. Potency period 5 years.
		Habu-venom Antivenin	*Trimeresurus flavoviridis* (Okinawa habu)	
	Laboratory of Chemotherapy and Serum Therapy, Kumamoto City, Kyushu	Antivenin Habu	*Trimeresurus flavoviridis* (Okinawa habu)	Concentrated and lyophilized.
	The Takeda Pharmaceutical Co., Osaka	Antivenin Mamushi	*Agkistrodon halys blomhoffii* (Mamushi)	Concentrated but not lyophilized.
MEXICO	Instituto Nacional de Higiene, Czda. M. Escobedo No. 20, Mexico 13, D. F.	Suero Anticrotálico	*Crotalus durissus* ssp. (Cascabel), *C. basiliscus* (Mexican west coast rattlesnake)	Ammonium sulfate precipitated and dialyzed. Reported to neutralize venoms of all Mexican crotalids.
		Suero Antibotrópico	*Bothrops atrox* (Barba amarilla)	
		Suero Antiviperino	*Crotalus durissus* ssp. (Cascabel), *C. basiliscus* (Mexican west coast rattlesnake), *Bothrops atrox* (Barba amarilla)	
	Laboratorios MYN, S.A., Av. Coyoacan 1707, Mexico 12, D.F.	Suero Anticrotálico MYN Liofilizado	Different Mexican *Crotalus*	All antivenins are modified by enzymes and concentrated. Suero Anticrotálico is lyophilized. Each vial neutralizes 15 mg. of *C. durissus* venom. Potency period is 5 years. Suero Antibothrópico is lyophilized. Each vial neutralizes 30 mg. of *B. atrox* venom. Potency period is 5 years. Suero Antiviperino is lyophilized. Each vial neutralizes 15 mg. of
		Suero Antibothrópico MYN Liofilizado	*Bothrops atrox* (Barba amarilla)	
		Suero Antiviperino MYN Liofilizado	Different Mexican *Crotalus* and *Bothrops atrox* (Barba amarilla)	
		Suero Antiofídico MYN Polivalente (Liquido)	Mexican *Crotalus* and *Bothrops atrox* (Barba amarilla)	

			C. durissus venom and 30 mg. of *B. atrox* venom. Potency period is 5 years. Suero Antiofídico is a liquid preparation. Each ampule of 10 ml. neutralizes 15 mg. of *C. durissus* venom and 30 mg. of *B. atrox* venom. Potency period is 2 years. It also contains 1500 I.U. tetanus antitoxin.
PHILIPPINE REPUBLIC-------	Bureau of Research and Laboratories, P.O. Box 911, Manila	Cobra Antivenin	*Naja n. philippinensis* (Philippine cobra)	Concentrated.
REPUBLIC OF SOUTH AFRICA----	South African Institute for Medical Research, Hospital Hill, P.O. Box 1038, Johannesburg	Boomslang Antivenin	*Dispholidus typus* (Boomslang)	All antivenins are pepsin treated, thermocoagulated, salt precipitated, and concentrated in liquid form. Boomslang antivenin is not ordinarily distributed but is available upon request. Potency period is 3 years (or longer if stored at 2-5° C.).
		Polyvalent Antivenin (Bitis, Naja, Hemachatus)	*Bitis arietans* (Puff adder), *B. gabonica* (Gaboon viper), *Naja nivea* (Yellow cobra), *Hemachatus haemachatus* (Ringhals)	Fitzsimons' Snake Park Laboratory, Snell Parade, Durban, no longer produces antivenins but acts as distributor for those produced by the South African Institute for Medical Research.
		Polyvalent Antivenin (Hemachatus, Naja, Bitis, Echis)	*Hemachatus haemachatus* (Ringhals), *Naja nivea* (Yellow cobra), *Bitis arietans* (Puff adder), *B. gabonica* (Gaboon viper), *Echis carinatus* (Saw-scaled viper)	
		Polyvalent Antivenin (Dendroaspis)	*Dendroaspis angusticeps* (Eastern green mamba), *D. jamesoni* (Jameson's mamba), *D. polylepis* (Black mamba)	

ANTIVENIN SOURCES (Con't.)

COUNTRY (Cont'd)	PRODUCER (Cont'd)	NAME OF PRODUCT (Cont'd)	VENOM(S) USED IN PREPARATION (Cont'd)	COMMENTS (Cont'd)
RHODESIA	CAPS, P.O. Box 2279, Salisbury	CAPS Snake Bite Antivenin	*Bitis arietans* (Puff adder, *Naja nivea* (Cape cobra), *Hemachatus haemachatus* (Ringhals), *Bitis gabonica* (Gaboon viper)	Reported to neutralize venom of *Dendroaspis angusticeps*. Supplied as globulin in liquid form.
TAIWAN	Taiwan Serum & Vaccine Laboratory, 151 Tong-shin Street, Nang Kang, Taipei	Bungarus Monovalent	*Bungarus multicinctus* (Many-banded krait)	*Agkistrodon* monovalent antivenin is reported to neutralize Chinese habu venom and hemorrhagic polyvalent antivenin is reported to neutralize *Agkistrodon* venom to a limited extent. These are liquid antivenins. Not concentrated. Potency period is 2 years. Supplied only to Taiwan addressees (1967).
		Naja Monovalent	*Naja n. atra* (Chinese cobra)	
		Agkistrodon Monovalent	*Agkistrodon acutus* (Sharp-nosed pit viper)	
		Neurotoxic Polyvalent	*Bungarus multicinctus* (Many-banded krait), *Naja n. atra* (Chinese cobra)	
		Hemorrhagic Polyvalent	*Trimeresurus mucrosquamatus* (Chinese habu), *T. stejnegeri* (Chinese green tree viper)	
THAILAND	Queen Saovabha Memorial Institute, Rama IV Street, Bangkok	Antivenine Serum "Banded Krait"	*Bungarus fasciatus* (Banded krait)	Lyophilized whole serums with a potency period of 5 years. One ml. of these antivenins will neutralize 0.8 mg. of the homologous venom, except the "Cobra" antivenin which will neutralize only 0.4 mg.
		Antivenine Serum "Cobra"	*Naja n. kaouthia* (Monocellate cobra)	
		Antivenine Serum "Green Pit Viper"	*Trimeresurus popeorum* (Pope's tree viper)	
		Antivenine Serum "King Cobra"	*Ophiophagus hannah* (King cobra)	
		Antivenine Serum "Russell's Viper"	*Vipera russelii* (Russell's viper)	
		Antivenine Serum "Pit Viper"	*Agkistrodon rhodostoma* (Malayan pit viper)	

	Product	Species	Notes
UNITED STATES —— Wyeth Laboratories, Inc, Marietta, Pa.	North and South American Crotalid Antivenin	*Crotalus adamanteus* (Eastern diamondback rattlesnake), *C. atrox* (Western diamondback rattlesnake), *C. durissus terrificus* (Cascabel), *Bothrops atrox* (Barba amarilla)	This antivenin neutralizes the venoms of most pit vipers. It is supplied as ammonium sulfate concentrated lyophilized globulins.
	Antivenin (Micrurus fulvius) (Coral Snake Antivenin)	*Micrurus fulvius fulvius* (Eastern coral snake)	This product, expected to be available in the near future, is a concentrated, refined, lyophilized horse serum antivenin, marketed in packages of 5 vials of antivenin and 5 vials of diluent (distilled water with preservative). The contents of each vial, when reconstituted with 10 ml. of diluent, will neutralize approximately 2.5 mg. of the homologous venom and approximately the same amount of the venom of *M. nigrocinctus* (Black-banded coral snake). It does not neutralize the venom of *M. mipartitus* (Black-ringed coral snake) or *Micruroides euryxanthus* (Arizona coral snake). Potency period is 5 years.
U.S.S.R. ——————— Tashkent Institute, Ministry of Health, Moscow	Cobra Antivenin	*Naja n. oxiana* (Oxus cobra)	The product names are only approximations of the Russian. No information is available as to the concentration or manner of preparation of these antivenins.
	Echis Antivenin	*Echis carinatus* (Saw-scaled viper)	
	Gyurza Antivenin	*Vipera lebetina* (Levantine viper)	
	Gyurza Polyvalent Antivenin	*Vipera lebetina* (Levantine viper), *Naja n. oxiana* (Oxus cobra)	
	Echis Polyvalent Antivenin	*Echis carinatus* (Saw-scaled viper), *Naja n. oxiana* (Oxus cobra)	

ANTIVENIN SOURCES (Con't.)

COUNTRY (Cont'd)	PRODUCT (Cont'd)	NAME OF PRODUCT (Cont'd)	VENOM(S) USED IN PREPARATION (Cont'd)	COMMENTS (Cont'd)
VENEZUELA	Laboratorio Behrens, Calle Real de Chapellin, Apartado 62, Caracas, D. F.	Suero Antibotrópico	*Bothrops atrox* (Barba amarilla), *Bothrops* sp. ("Tigra Mariposa")	The name "Tigra Mariposa" is applied to at least two species of *Bothrops* in Venezuela, and it is not clear which species' venom is used in preparation of the antivenin. These antivenins are treated to reduce the amount of foreign protein.
		Suero Anticrotálico	*Crotalus durissus* ssp. (Cascabel)	
		Suero Antiofídico Polivalente	*Bothrops atrox* (Barba amarilla), *Bothrops* sp. ("Tigra Mariposa"), *Crotalus durissus* ssp. (Cascabel)	
VIET NAM	Institut Pasteur Nha Trang	Sea Snake Antivenin[2]	*Lapemis hardwickii* (Hardwicke's sea snake)	Serum not fractionated or or concentrated.
YUGOSLAVIA	Institute for Immunology, Rocke-fellerova 2, Zagreb	Serum Antiviperinum	*Vipera ammodytes* (Long-nosed viper)	Serum digested with pepsin and precipitated with ammonium sulfate.

[1] Included when available.
[2] Production temporarily suspended until venom can again be obtained.

GLOSSARY

Anal plate: The large scute covering the vent. It marks the division between body and tail. It may be *entire*, or *divided* by an oblique suture (fig. 10).

Anaphylaxis: A severe hypersensitivity reaction which may cause circulatory, respiratory and neurological symptoms. Often fatal if untreated.

Antivenin: An antitoxic serum which neutralizes a venom.

Antivenom: Antivenin.

Apical pits: Tiny depressions, usually paired, near the terminal end of each dorsal scale when present; function unknown.

Aquatic: Living in water. (Compare terrestrial.)

Arboreal: Living in trees or bushes. (Compare terrestrial.)

Autopharmacological substances: Chemicals produced and released by body cells in response to a stimulus, such as venom. These substances may produce deleterious effects, such as shortness of breath, changes in heart rate and shock.

Canaliculated: Traversed by a small tubular passage or channel. Here applied to the fangs of snakes.

Canthus (or canthus rostralis): The angle between the flat crown of the head and the side, between snout and eye; may be *sharp*, *obtuse*, *obsolete*, or *absent*.

Canthal scales: Enlarged scales along the lateral border of the crown between internasals and supraoculars in some vipers and pit vipers (fig. 12).

Chin shields (Genials): Paired enlarged scales near the ventral midline of the lower jaw; *anterior chin* shields in contact with mental or separated from it by first infralabials; a pair of *posterior chin* shields may be present behind the anterior (fig. 6).

Compressed: (In reference to body shape.) Flattened from side to side, giving a greater height than breadth.

Constriction Band: A wide piece of rubber or other material used to depress flow along superficial lymphatic and venous channels.

Crotalid(s): Refers to snakes of the family Crotalidae or pit vipers which includes the rattlesnakes and lance-heads.

Crown: The top of the head, or the anterior part of the top; usually occupied by 9 enlarged scutes, from the rostral (on the snout) to the parietals (fig. 6).

Cyanosis: Bluish discoloration of the skin caused by insufficient oxygenation of the blood.

Depressed: (In reference to body shape.) Flattened from top to bottom (dorsoventrally), giving a greater breadth than height.

Distal: Farther away from the body. (Compare proximal.)

Diurnal: Active during the daylight hours (see nocturnal).

Dorsals: The rows of small scales that cover the top (dorsal) surface of a snake's body. They are counted in a diagonal (or zigzag) line from the edge of the ventral plate, across the back to the opposite edge (fig. 7).

Ecchymosis: A discoloration of the skin resembling a bruise. It is caused by the extravasation of blood.

Edema: The presence of excessive fluid in the intercellular tissue spaces.

Elapid(s): Refers to the snakes of the family Elapidae which includes the cobras, kraits, coral snakes, and mambas.

Envenomation: The deposition of venom within tissues.

Extravasation: Passing of a body fluid out of its proper place, as blood into surrounding tissues after rupture of a vessel.

Eye (Sizes): An eye of *moderate* size has a diameter that is about equal to the eye's distance from the lip; a *large* eye's diameter is about one and one-half this distance; a *small* eye is about one-half this distance.

Fangs: Enlarged hollow or grooved teeth specialized for injection of venom (fig. 1).

Fasciotomy: An incision cutting the fascia or dense connective tissue that surrounds muscles. Sometimes used in treatment of snakebite to release pressure from severe swelling.

Frontal: The single enlarged median scute on the crown between the supraoculars and behind the prefrontals (fig. 6).

Gulars: The rows of small nonspecialized scales on the ventral surface of the lower jaw anterior to the ventral plates (fig. 6).

Hemoglobinuria: The presence of hemoglobin in the urine.

Hemotoxin: A toxin capable of destroying blood cells. Often also applied to toxins that cause hemorrhage.

Herpetology: The scientific study of reptiles and amphibians.

Hypotension: Abnormally low blood pressure.

Imbricate: Overlapping, as the tiles on a roof; the usual condition for dorsal and ventral scales. (Compare juxtaposed.)

Infralabials: The (usually enlarged) scales along the border of the lower lip behind the mental (fig. 6).

Internasal(s): The (usually paired) scutes on the crown just behind the rostral (fig. 6).

Intraperitoneal: Within the peritoneum, or peritoneal cavity, as intraperitoneal injection of drugs.

Juxtaposed: With edges adjacent but not overlapping; the usual condition for head shields. (Compare imbricate.)

LD$_{50}$: The amount of a drug or poison necessary to kill 50 percent of the animals in a test group; usually stated in mg. per kg. on a dry basis.

Loreal: The scale on the side of the head lying between the (post-) nasal and the preocular(s); characteristically *absent* in elapid snakes (fig. 6).

Lyophilization: Process of quick freezing and dehydration under a high vacuum.

Mental: The triangular scale at the symphysis of the lower jaw, corresponding in position to the rostral of the upper jaw (fig. 6).

Myoglobinuria: Presence of a type of muscle protein in the urine.

Nasal: The scale enclosing the nostril; may be *single, partially divided* (by a suture extending down from the nostril), or *completely divided* (giving a pre- and post-nasal) by a vertical suture that extends through the nostril (fig. 6).

Nasal valve: A sphincter device for closing the nostrils; found in some vipers and nearly all sea snakes.

Nasorostral (scale): An enlarged scale (usually paired) that lies just behind the rostral scale (e.g., between the rostral and the nasal) in some vipers.

Necrosis: Death of a circumscribed portion of tissue.

Necrotizing: Capable of causing necrosis.

Neuromuscular transmission: The relay of a stimulus from the end of a nerve to its muscle.

Neurotoxin: A poison that has a more marked effect on nerve tissue than other body tissues. Often improperly used to denote that the poison affects *only* the nervous system.

Neutralize: The ability of a substance (antivenin) to nullify the effects of another substance (venom).

Nocturnal: Active during the night. (Compare diurnal).

Occipitals: Paired enlarged scutes that lie immediately behind the parietals in a few snakes, e.g. king cobra (fig. 81).

Paresis: A slight or incomplete paralysis, sometimes noted as a weakness of a muscle.

Paresthesia: An abnormal sensation, as pricking, numbness or burning.

Parietals: The large paired scutes at the rear end of the crown, immediately behind the frontal and supraoculars (fig. 6).

Petechiae: Small spots formed by the effusion of blood.

Pit (or Loreal Pit): The deep depressions on the side of the head in the loreal region in pit vipers (family Crotalidae); they are heat-sensitive and aid the snake in finding its prey in the dark (fig. 4).

Plate: A large flat scale, usually on the head or ventral surface.

Polyvalent: Used in this text to denote a serum

containing antitoxins against the venoms of a number of different snakes.

Postoculars: The scale(s) immediately behind and in contact with the eye. Usually between the supralabials and the supraoculars (fig. 6).

Prefrontal(s): The (usually paired) enlarged scutes just behind the internasals, or that area if it is covered with small scales (fig. 6).

Prehensile (in reference to tail): Adapted for grasping by wrapping around; usually visible as a curled and compressed tail tip. Typical of tree snakes.

Preoculars: The scale(s) lying immediately in front of and in contact with the eye (fig. 6).

Proteolytic: Capable of causing the digestion or dissolution of proteins.

Proximal: Nearest to the main part of the body or the median line of the body. (Compare distal.)

Ptosis: A drooping of the upper eyelid.

Pupil (of eye): The black opening enclosed by the iris; may be round, horizontally elliptical, or vertically elliptical.

Rostral: The single enlarged plate at the tip of the snout (fig. 6).

Savannah: Open grassy country interspersed with small groups of trees or bushes.

Scutes: Overlapping or juxtaposed scales that cover the surface of the body. Formed of horny skin in reptiles, differing from the bony scales of fishes.

Shield: An enlarged scale or scute, commonly specialized and with a distinctive name.

Shock position: Victim lying on his back with head and chest slightly lower than his feet.

Snakebite: A bite inflicted by either a venomous or nonvenomous snake.

Snake venom poisoning: A condition resulting from the injection of snake venom.

Subcaudals: The scales or scutes under the tail; they may extend across the entire ventral surface (single), or go only half way across, where they are met by another scale (paired) (fig. 10).

Suboculars: The scale(s) immediately below and in contact with the eye; between the eye and the supralabials (fig. 6).

Subterranean: Living under the surface of the ground. (Compare terrestrial.)

Supra-anal tubercles: Small raised keel-like structures on the dorsal scales above the vent in some snakes.

Supralabials: The scales (usually enlarged) or scutes along the border of the upper lip behind the rostral (fig. 6).

Supraoculars: The enlarged scales or scutes (sometimes divided) on the crown directly above each eye (fig. 6).

Suture: A line of division between two scales.

Swelling: An enlarged area.

Temporals: Scales or scutes on the side of the head between the parietals and the supralabials, and behind the postoculars; anterior temporal(s) are in contact with the postoculars; posterior (sometimes secondary and tertiary) temporals are in vertical rows, not in contact with postoculars (fig. 6).

Terrestrial: Living on land or on the ground. (See aquatic, arboreal, subterranean.)

Toxin: A naturally occurring poisonous substance. A synonym for venom or poison.

Trismus: Tetanic spasm of jaw muscles; lockjaw.

Urticaria: A skin eruption, usually associated with allergy, characterized by sudden appearance of smooth, slightly elevated patches usually paler than the surrounding skin and accompanied by itching. Commonly called hives or nettle rash.

Vasopressor: A drug that raises blood pressure by stimulating the contracting muscles of the capillaries and arterioles.

Veldt: The open grassy regions of the African highlands.

Venom apparatus: The structural components that produce, transport and deliver the venom. In snakes, it is usually composed of two venom glands, two venom ducts, and two or more teeth or fangs.

Vent: The common posterior opening of the urinary, gastrointestinal, and reproductive systems of reptiles; marks the beginning of the tail in snakes.

Ventrals: The enlarged scales (scutes or plates) that extend down the undersurface of the body (fig. 9).

Vesiculation: The formation of blisters.

GENERAL REFERENCES

BOGERT, C. M. 1943. Dentitional Phenomena in Cobras and Other Elapids with Notes on Adaptive Modifications of Fangs. Bull. Amer. Mus. Nat. Hist., vol 81 (art 3), pp. 285–360, figs. 1–73, pls. 48–51, maps 1–4.

BOULENGER, G. A. 1896. Catalogue of the Snakes in the British Museum (Natural History). Vol III. Taylor and Francis, London, xiv + 727 pp., 37 figs., 25 pls.

BRATTSTROM, B. H. 1964. Evolution of the Pit Vipers. Trans. San Diego Soc. Nat. Hist., vol 13, pp. 185–268, figs. 1–41.

BUCKLEY, E. E. and PORGES, N. (editors) 1956. Venoms. Pub. Amer. Assoc. Advancement Sci. No. 44, xii + 467 pp., text figs.

COCHRAN, D. M. 1943. Poisonous Reptiles of the World: a Wartime Handbook. Smithsonian Inst. War Background Studies No. 10, v + 37 pp., 2 figs., 17 pls. (1 color).

DITMARS, R. L. 1931. Snakes of the World. Macmillan, New York, 207 pp., 84 pls. (Reprinted in paperback, 1962, Pyramid, New York).

GLOYD, H. K. and CONANT, R. 1943. A Synopsis of the American Forms of *Agkistrodon* (Copperheads and Moccasins). Bull. Chicago Acad. Sci., vol 7, pp. 147–170, figs. 1–16, maps 1–2.

KAISER, E. and MICHL, M. 1958. Die Biochemie der Tierischen Gifte. F. Deuticke, Wien, 258 pp.

KEEGAN, H. L. and MACFARLANE, W. V. (editors) 1963. Venomous and Poisonous Animals and Noxious Plants of the Pacific Region. Pergamon Press, Oxford, 456 pp., text figs.

KLAUBER, L. M. 1956. Rattlesnakes: Their Habits, Life Histories and Influence on Mankind. Univ. California Press, Berkeley, 2 vols., 1476 pp., 187 figs. (2 color).

KLEMMER, K. 1963. Liste der Rezenten Giftschlangen: Elapidae, Hydrophidae, Viperidae und Crotalidae. In, Die Giftschlangen der Erde. N. G. Elwert, Marburn/Lahn, pp. 254–464, pls. 1–37 (color), map 1.

LUDICKE, M. 1962, 1964. Serpentes. In, Handbuch der Zoologie. Walter de Gruyter and Co., Berlin, vol 7, part 1, fasc. 5 and 6, 298 pp., illus.

MAKI, M. 1931. A Monograph on Snakes of Japan. Dai-ichi Shobo (publisher), Tokyo, Japan. 240 pp.

MARX, H. and RABB, G. B. 1965. Relationships and Zoogeography of the Viperine Snakes (Family Viperidae). Fieldiana Zool., vol 44, pp. 161–206, figs. 32–46.

MASLIN, T. P. 1942. Evidence for the Separation of the Crotalid Genera *Trimeresurus* and *Bothrops* with a Key to the Genus *Trimeresurus*. Copeia, No. 1, pp. 18–24, figs. 1–2.

MERTENS, R. 1960. The World of Amphibians and Reptiles. (English translation by H. W. Parker). George G. Harrap and Co., London, 207 pp., 80 pls. (16 color).

NATIONAL ACADEMY OF SCIENCES, NATIONAL RESEARCH COUNCIL, COMMITTEE ON SNAKEBITE THERAPY 1963. Interim Statement on Snakebite Therapy. Toxicon, vol 1, pp. 81–87.

OLIVER, J. A. 1959. Snakes in Fact and Fiction. Macmillan, New York, xiii + 199 pp., illus.

PARKER, H. W. 1963. Snakes. W. W. Norton Co., New York, 191 pp., illus.

PHISALIX, M. 1922. Animaux Venimeux et Venins. Masson, Paris, vol II, xii + 864 pp., 521 figs., 17 pls. (8 color).

POPE, C. H. 1955. The Reptile World. Alfred A. Knopf, New York, xxv + 324 pp., 221 figs.

RUSSELL, F. E. and SCHARFFENBERG,

R. S. 1964.   Bibliography of Snake Venoms and Venomous Snakes. Bibliographic Assoc., West Covina, California, 220 pp.

SCHMIDT, K. P. and INGER, R. F. 1957. Living Reptiles of the World. Hanover House, Garden City, New York, 287 pp., text figs., 145 color pls.

SWAROOP, S. and GRAB, B. 1954.   Snakebite Mortality in the World.   Bull. World Health Org., vol 10, pp. 35–76.

NOTES

INDEX

This index has been prepared to help the layman seeking information about poisonous snakes. The technical names of all the poisonous snakes are listed alphabetically with the species name first and the genus name following it. The most widely used common names of the snakes are also listed alphabetically, followed by the technical names in their customary order (genus name first). Since common names are sometimes misleading, identification based on common names is not recommended, and all information about the snakes will be found under the technical names only. The index also contains the major subjects treated in the text and the key words most helpful to the reader seeking specific information. Maps and geographical distribution tables are also cross-indexed here.

Common names in foreign languages have not been listed according to the rules of alphabetizing in those languages, but have been treated as if they were English words in order to make it easier for the English-speaking layman to find the information he needs quickly.

192

196

201